Springer Theses

Recognizing Outstanding Ph.D. Research

Aims and Scope

The series "Springer Theses" brings together a selection of the very best Ph.D. theses from around the world and across the physical sciences. Nominated and endorsed by two recognized specialists, each published volume has been selected for its scientific excellence and the high impact of its contents for the pertinent field of research. For greater accessibility to non-specialists, the published versions include an extended introduction, as well as a foreword by the student's supervisor explaining the special relevance of the work for the field. As a whole, the series will provide a valuable resource both for newcomers to the research fields described, and for other scientists seeking detailed background information on special questions. Finally, it provides an accredited documentation of the valuable contributions made by today's younger generation of scientists.

Theses are accepted into the series by invited nomination only and must fulfill all of the following criteria

- They must be written in good English.
- The topic should fall within the confines of Chemistry, Physics, Earth Sciences, Engineering and related interdisciplinary fields such as Materials, Nanoscience, Chemical Engineering, Complex Systems and Biophysics.
- The work reported in the thesis must represent a significant scientific advance.
- If the thesis includes previously published material, permission to reproduce this must be gained from the respective copyright holder.
- They must have been examined and passed during the 12 months prior to nomination.
- Each thesis should include a foreword by the supervisor outlining the significance of its content.
- The theses should have a clearly defined structure including an introduction accessible to scientists not expert in that particular field.

More information about this series at http://www.springer.com/series/8790

Mengqi Fu

Electrical Properties of Indium Arsenide Nanowires and Their Field-Effect Transistors

Doctoral Thesis accepted by
the Peking University, Beijing, China

Springer

Author
Dr. Mengqi Fu
Department of Physics
University of Konstanz
Konstanz, Germany

Supervisor
Prof. Qing Chen
Key Laboratory for the Physics
and Chemistry of Nanodevices
Department of Electronics
Peking University
Beijing, P. R. China

ISSN 2190-5053 ISSN 2190-5061 (electronic)
Springer Theses
ISBN 978-981-13-3443-6 ISBN 978-981-13-3444-3 (eBook)
https://doi.org/10.1007/978-981-13-3444-3

Library of Congress Control Number: 2018962397

This Springer imprint is published by the registered company Springer Nature Singapore Pte Ltd.
The registered company address is: 152 Beach Road, #21-01/04 Gateway East, Singapore 189721, Singapore

Supervisor's Foreword

Si-based semiconductor technology and industry have been developed very successfully in the last decades and have changed our lives enormously. One of the most successful routes to develop Si-based field-effect-transistors (FETs), the basic devices for logic computing and amplification, is scaling down in which the geometric sizes of the devices are reduced proportionally and the performances of the devices are improved. However, in the new century, such successful developing route faces difficulties as the thickness of the SiO_2 dielectric layer has been reduced to a couple of nanometers, and the diameter and length of the conducting channel have been reduced to the scale of 10 nm. New channel materials with higher mobility than Si are considered to solve the difficulties in addition to introducing high-K dielectric materials and Fin- or gate-all-around structure. InAs is a typical III–V material with high electron mobility. The electron mobility of bulk InAs is 40,000 cm^2/V s, which is much higher than 1400 cm^2/V s of Si. In addition, InAs also has small bandgap (0.35 eV), high electron injection speed, and easiness to form ohmic contact with metals. However, it has been reported that when the diameter of InAs nanowires (NWs) decreases continuously to 15 nm, the electron mobility decreases linearly to about 2500 cm^2/V s. Another issue is that when the diameter of InAs NWs decreases, the stable phase changes from zinc blende (ZB) phase to wurtzite (WZ) phase and the properties of WZ phase InAs are not very clear. In this thesis, Dr. Mengqi Fu has first studied the performance of FETs based on WZ phase thin InAs NWs and revealed their change with the NW diameter. Next, she has developed a novel method which allows characterizing the NWs in FET by transmission electron microscopy (TEM). Such method allows direct correlating the performance of FET with the atomic-level structure of the channel NW. With this method, Dr. Fu has studied the effects of phase and crystal orientation to the properties of InAs NWs and their FETs.

In her first work, in order to exclude other effects, she has chosen InAs NWs grown by molecular beam epitaxy (MBE) with pure WZ phase and along <0001> axial direction. The thinnest diameter of the NWs she has used is 7 nm, which is the thinnest InAs NW having been used in FETs at that time and is the same size of the channel diameter in 7 nm node FETs. She has revealed quantitatively that all the key parameters of the InAs NW FETs change with the NW diameter. Her results are significant for further development of semiconductor technology.

Not only the diameter, but also the phase and the orientation of the channel material can change the properties of the NWs and their devices. TEM is the most effective technique to characterize atomic-level structure of nanomaterials. However, the as-grown NWs normally have different structures from one to another. How to characterize the structure of the specific NW in FET is a big challenge. Dr. Mengqi Fu has developed a novel method based on a special device fabrication process and nano-manipulation. With her method, one can suspend the NWs in FETs after electrical measurements, and can then move the NWs into TEM for structure characterization. This method provides a technique to correlate the performance of devices (not limited to be FETs) with the atomic-level structure of the key nanostructures in the devices.

Using the novel method she has developed, Dr. Mengqi Fu has found for the first time how the crystal phase, WZ or ZB, and the orientation of InAs NWs remarkably affect the electronic properties of InAs NWs and their FETs. She has also studied the different electronic properties of InAs NWs grown by MBE and metal-organic chemical vapor deposition (MOCVD). These results broaden our knowledge on the electronic properties of nanomaterials and their devices, provide significant reference for further developing nanodevices, and are also significant for understanding the related physical and electronic phenomenon.

Beijing, P. R. China
November 2018

Prof. Qing Chen

Abstract

As the rapid development of the semiconductor industry, modern silicon-based integrated circuits (IC) technology has come to the 14 nm node, approaching the physical and technological limit of silicon material. To promote the IC technology, it is needed to introduce novel materials and device structures. The III–V compound semiconductors, such as indium arsenide (InAs) and gallium arsenide, have demonstrated their great potential as the alternative material for the next generation of high-speed electronic devices, due to their ultrahigh carrier mobility. On the other hand, since the gate-all-around nanowire (NW) devices can efficiently suppress the short-channel effects owing to their favorable electrostatic geometry, the III–V compound semiconductor NWs are attractive building blocks for nanoelectronics. In this thesis, the electrical transport properties of individual InAs NW-based field-effect transistor (FET) are studied. By studying the influence of crystal structure, size, and growth method on electrical transport properties of InAs NWs, the optimizations to achieve the better performance InAs NW electronic devices are explored.

First, the high-performance FETs are fabricated based on the single-crystalline wurtzite (WZ) ultrathin InAs NWs grown along <0001> direction. By studying the key parameters of devices, the opportunities and challenges that are brought by scaling down InAs NWs are presented. When the diameter of InAs NW is smaller than 10 nm, as the NW diameter decreases, the devices have larger ON-OFF ratios, higher resistivity, and exhibited a positive shift in threshold voltage due to the more efficient gate control capability and quantum confinement effects. By using the thinnest ($\sim$7 nm) NW, the high ON-OFF ratio up to 2×10^8 is obtained and is higher than all the value in all the reported InAs NW FETs. A fall in the electron field-effect mobility is also seen due to the surface roughness scattering and lower density of states in ultrathin NWs. Moreover, the normalized contact resistance drastically increases within the smaller diameter InAs NWs; even no barrier is found between the NWs and metal electrodes.

Next, the influence of crystal phase and crystal orientation on the electrical transport properties is studied by developing a novel fabrication process and a particular characterization method. To correlate the electrical properties and crystal structures of the same InAs NW, a technique is developed by suspending the InAs NW devices after electrical measurements and then transferring the NWs into TEM by the nano-manipulation. With this technique, it is observed that the crystal phase, WZ or ZB, and the orientation of the InAs NWs remarkably affects the electronic properties of NW FETs, such as the threshold voltage, ON-OFF ratio, subthreshold swing, and effective barrier height at OFF-state. The WZ InAs NWs have obvious smaller field-effect mobility, conductivities, and electron concentration at $V_{BG} = 0V$ than the ZB InAs NWs, while these parameters are not sensitive to the orientation of the ZB InAs NWs. The subthreshold swing increases while threshold voltage, ON-OFF ratio, and effective barrier height at the OFF-state decrease in the sequence of WZ <0001>, ZB <131>, ZB <332>, ZB <121>, and ZB <011>. The good ohmic contact between InAs NWs and metal remains regardless of the variation of the crystal phase and orientation.

Finally, the differences on electrical properties of InAs NWs grown by the two mostly used growth methods, metal-organic chemical vapor deposition (MOCVD) and molecular beam epitaxy (MBE), are studied. The NWs grown by two methods both show the larger ON-OFF ratio when the NWs become thinner. But the relation between diameter and ON-OFF ratio is different. The ON-OFF ratio of thin NWs (20 nm thickness) is similar in MBE-grown and most MOCVD-grown NWs. However, the MOCVD-grown NWs show a much more diversity on the OFF-state current than the MBE-grown NWs. When the diameter of InAs NWs becomes larger (about 100 nm), the ON-state performance of two kinds of NWs is similar but the OFF-state performance of MBE-grown NWs is much better than that of the MOCVD-grown NWs. The results of the finite elements simulation confirm that the different doping levels caused by background carbon dopant in MOCVD-grown InAs NWs might be the main factor leading to the difference of electrical properties between MOCVD- and MBE-grown InAs NWs.

Keywords InAs nanowire · Field-effect transistor · Electrical properties
Ultrathin nanowire · Crystal orientation · Crystal phase · Growth methods

Parts of this thesis have been published in the following journal articles:

1. "Crystal phase- and orientation-dependent electrical transport properties of InAs nanowires," **M. Q. Fu**, Z. Q. Tang, X. Li, Z. Y. Ning, D. Pan, J. H. Zhao, X. L. Wei, and Q. Chen, *Nano Lett.*, 16, 2478–2484 (2016).
2. "Electrical characteristics of field-effect transistors based on indium arsenide nanowire thinner than 10 nm," **M. Q. Fu**, D. Pan, Y. J. Yang, T. W. Shi, Z. Y. Zhang, J. H. Zhao, H. Q. Xu, and Q. Chen, *Appl. Phys. Lett.*, 105, 143101 (2014).
3. "The structure-dependent properties of InAs nanowires and their devices," **M. Q. Fu**, T. W. Shi, X. Li and Q. Chen, *ECS Trans.,* 75(8), 733–748 (2016).

Acknowledgements

I owe my sincere gratitude to my supervisor Prof. Qing Chen for accepting me as a Ph.D. student at Peking University. Thank you for your continuous encouragement and support for my research work and for the invaluable advice of my future career. I couldn't reach this far without the efforts and time you have spent on me. I have truly enjoyed all the discussions and all the critical comments, which have become a great treasure and memory for me.

I'd also like to thank Prof. Hongqi Xu, Prof. Shengyong Xu, Prof. Zhiyong Zhang, Prof. Xianlong Wei, and Prof. Ning Kang for providing many insightful comments on the manuscript and for many inspiring discussions during the project. I would also like to acknowledge Prof. Jianhua Zhao, Prof. Tao Yang, Dr. Dong Pan, and Dr. Xiaoye Wang at the Institute of Semiconductor, Chinese Academy of Science, for providing InAs nanowire samples, without which this work could not be finished.

I'd like to thank Dr. Tuanwei Shi, Dr. Zhiyuan Ning, Xing Li, Zhiqiang Tang and Yuxiang Han for being great colleagues and friends to me.

Last but not least, I would like to thank my parents and my husband Dingxun Fan for your love and support throughout the long and winding journey.

Cologne, Germany
October 2018

Mengqi Fu

Contents

Chapter 1
Introduction

Abstract As the miniaturization and integration of solid-state electronic devices has continued to increase rapidly with the demands of high speed, low power consumption and high storage density, the conventional Si-based technology has lost their advantages on fabrication process. Therefore the technologies based on new materials gradually attract researchers' attention. Among them, Indium Arsenide (InAs) nanowires (NWs) with high electron mobility is one of the most promising candidate. In this chapter, we introduce the advantages of InAs nanowire on electronic devices and the development status of InAs nanowire electronic devices. Also, the topic ideas and chapter arrangements of this thesis are presented.

Since the advent of integrated circuits in the late 1950s, the miniaturization and integration of solid-state electronic devices has continued to increase rapidly with the demands of high speed, low power consumption, and high storage density. In 1965, one of Intel's co-founders, Gordon Moore, proposed that "the number of transistors per square inch on integrated circuits will be doubled every 18–24 months, and meanwhile their performance will also be improved". This later became the so-called *Moore's Law* which is still governing the development of the traditional microelectronics industry, and the number of field-effect transistors (FETs) on a single chip has expanded from several thousand in 1970 to several billions at present. With the rapid development of modern semiconductor industry, silicon (Si)-based microelectronics process has reached the 14 nm node at the time of writing this thesis, which is nearing its technical and physical limits. Meanwhile, the reduction in the scale of devices has also lead to many problems, such as deviation from classic Boltzman diffusion model, excessive power consumption, increase of the source-drain contact resistance, short channel effects, and carrier speed saturation. To solve these problems, on the one hand, researchers are trying to optimize the fabrication process of Si devices and develop novel device structures (such as Silicon-on-Insulator (SOI), Fin-FET and Tri-gate FET). On the other hand, in order to prevent the increase of tunneling current caused by ultrathin gate oxide layer, silicon dioxide (SiO_2) is replaced by high-k dielectric as gate dielectric material after the 22 nm node. Nevertheless, this

M. Fu, *Electrical Properties of Indium Arsenide Nanowires and Their Field-Effect Transistors*, Springer Theses, https://doi.org/10.1007/978-981-13-3444-3_1

change on the fabrication process of results in the fact that a big advantage of Si technology, which is an ideal interface between Si and SiO_2, no longer exists. Therefore, researchers are gradually turning their attention to devices based on new materials and new working principles. Among them, Indium Arsenide (InAs) nanowires (NWs) with high electron mobility is one of the most promising candidate for reasons that will be discussed extensively in this chapter.

1.1 Advantages of InAs Nanowire-Based Electronic Devices

1.1.1 Advantages of InAs Material for Fabricating High-Performance Electronic Devices

As a basic component of a logic unit in integrated circuits, FETs can be classified into two types, the electron-conducting n-type FET and the hole-conducting p-type FET. As shown in Fig. 1.1, the FET switches at high speed between "ON" and "OFF" states [1]. Two p-type and n-type FETs can be combined to form the basic unit of the most widely used low-power, high-speed digital integrated circuit, which is called complementary metal oxide semiconductor (CMOS) transistor. The simple structure and its low power consumption characteristic of CMOS transistors is one of the foundations of large-scale integrated circuits. High speed FETs need to have a sufficiently high ON-state current (I_{on}) and transconductance (g) to reduce the carrier transport time in the channel. When the device is in the saturation region of I–V characteristics under operating voltage (V_{DD}), I_{on} and g are related to the surface density, mobility (μ), injection velocity (v_{inj}) and saturation velocity (v_s) of carriers in devices. Therefore, using materials having high mobility and carrier velocity is greatly beneficial for boosting the speed of FET.

It is also important to reduce the power consumption while increasing the speed of the device. At present, the power density of logic circuits has reached 100 W/cm^2 due to the scaling down of the devices. The energy (E_{total}) consumed by a CMOS device each time switching between "0" and "1" can be expressed as following:

$$E_{total} = E_{dynamic} + E_{leakage} = \alpha L_d C V_{DD}^2 + L_d I_{off} V_{DD} \tau_{delay} \quad (1.1)$$

with $E_{dynamic}$ the dynamic power, $E_{leakage}$ the static power, L_d the device logic depth, C the switched capacitance, τ_{delay} the delay time, α the logic activity factor, and I_{off} the OFF-state current. If the density of FETs further increases without reducing their V_{DD} and I_{off}, the power density and temperature of the chip will continue to rise, finally leading to the failure of FETs. In order to satisfy the requirements for normal application, a more expensive cooling power is needed, which will significantly increase the cost of production. Therefore, to further increase the integration density, it is desired to reduce the operating voltage of the FET while maintaining a low level of I_{off} [2–4]. However, as shown in Fig. 1.1, this will come at the expense of reducing

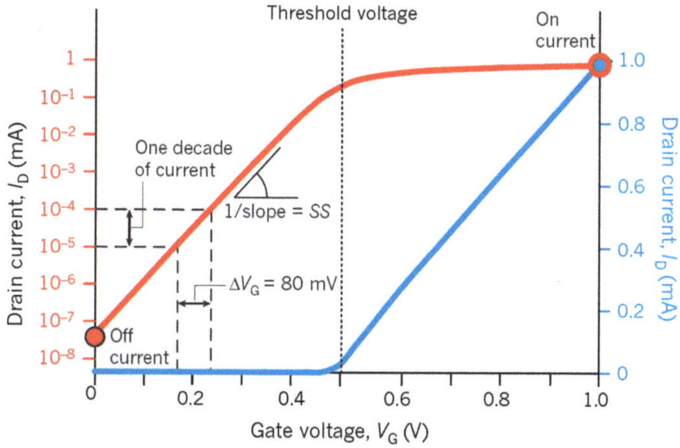

Fig. 1.1 The diagram of relation between source-drain current and gate voltage of a metal-oxide-semiconductor FET (MOSFET) device, which is also referred as the transfer curve (or transfer characteristics) of device. Wherein, the blue curve and the red curve are the transfer curves of device in linear and exponential coordinates, respectively. Reprinted from Ref. [1], with kind permission from Springer Nature

the ON-state current (I_{on}) and g of devices. In order not to lose the switching speed of FETs, the V_{DD} of FETs is maintained at around 0.8–0.9 V for the latest generations of products, [3, 5] which deviates from the constant electric field proportionality requirement to be followed for scaling down. If V_{DD} cannot be further reduced, scaling down of devices will not continue. In order to achieve a win–win situation of power consumption reduction and speed improvement, researchers are exploring a variety of new materials and new device structures. Introducing the III–V semiconductors which have higher carrier mobility and injection speed than Si as channel materials is one of the possible solutions. As shown in Table 1.1 and Fig. 1.2, many III–V materials (such as GaAs, InAs, and InSb) have higher electron mobility and electron injection velocity than Si. In particular, when the surface carrier density is similar, InGaAs or InAs materials have an order higher electron mobility than that of Si. Moreover, v_{inj} of electrons at $V_{DD} = 0.5$ V can reach 4×10^7 cm/s [6], which is more than twice of Si at $V_{DD} = 1.0$ V [7]. Higher drift velocity (v_{drift}) allows the device to obtain a faster response. In high-frequency electronic devices, the f_T of high electron mobility transistors (HEMTs) based on III–V semiconductors has reached several hundred GHz or even above THz [8, 9].

InAs is an important member of the high electron mobility III–V semiconductor compound family. Its electron mobility at room temperature can be up to 40,000 cm^2/V s. In reported works, the highest electron mobility of InAs materials in short-channel devices can reach 13,000 cm^2/V s, and their electron injection speed is also much larger than that of Si materials. Therefore, the current density of the HEMT based on two-dimensional thin film of III–V semiconductor using InAs as the channel material is much higher than that of Si-based device having the same chan-

Table 1.1 List of electrical parameters for several commonly used semiconductor materials

	Si	Ge	GaAs	InAs	InSb
m_{eff} (m_0)	0.19	0.08	0.067	0.023	0.014
μ_e (cm^2/V s)	1600	3900	9200	40,000	77,000
E_g (eV)	1.12	0.66	1.42	0.36	0.17
ε_r	11.8	16	12.4	14.8	17.7

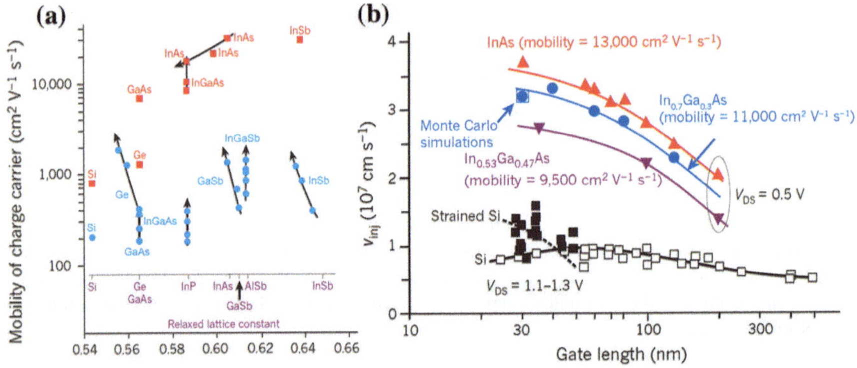

Fig. 1.2 **a** Diagram of the relation between reported highest mobility values of the electrons (red square data points) and holes (blue circular data points) in the inversion layer or quantum dots of devices based on semiconductor materials and the actual lattice constants of these semiconductors; **b** electron injection speed of different semiconductors in HEMT devices with different channel lengths. Reprinted from Ref. [10], with kind permission from Springer Nature

nel length (as shown in Fig. 1.3). On the other hand, as shown in Fig. 1.2a, because InAs and other III–V materials have similar lattice constants and similar crystal structures, by precisely controlling the growth conditions, researchers can design the energy band structure of the required materials based on desirable performance requirements, and epitaxially grow multilayer III–V materials with heterojunction structures [8, 11–14]. This flexible and controllable material system provides an excellent platform for studying nanostructures with new structures and principles.

1.1.2 Advantages of InAs Nanowires in Making Multi-gate and Gate-All-Around Devices

HEMT devices with InAs as channel materials have shown high performance in high speed and high frequency applications. However, because existing HEMT devices have no insulating layers, there is often serious gate leakage resulting in a high I_{off} [8]. The increase of I_{off} greatly raises the power consumption of the devices, and thus these devices cannot be applied to logic circuits with high integration density. To

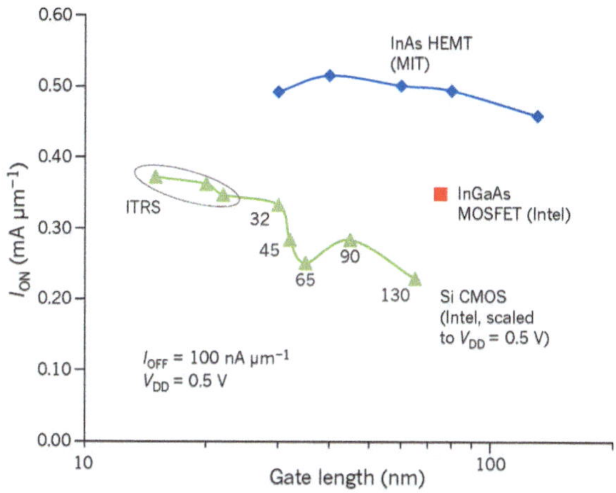

Fig. 1.3 Comparison of the ON-state current between InAs material-based HEMT devices and Si-based CMOS devices. As can be seen from the figure, when channel length is same, the InAs material-based devices has much larger ON-state current than the Si-based FET. Reprinted from Ref. [10], with kind permission from Springer Nature

this regard, III–V semiconductor MOSFETs or tunnel FETs (TFETs) with lower I_{off} levels have attacted intensive research interests. At the same time, in order to reduce the impact of short-channel effects (SCE) (e.g. increase of subthreshold swing (SS), drift of threshold voltage (V_T) and drain-induced barrier lowering (DIBL), which are induced by scaling down) on performance of devices (as shown in Fig. 1.4), the researchers have proposed that the existing planar HEMT structure can be replaced with multi-gate or even gate-all-around structure (Fig. 1.5).

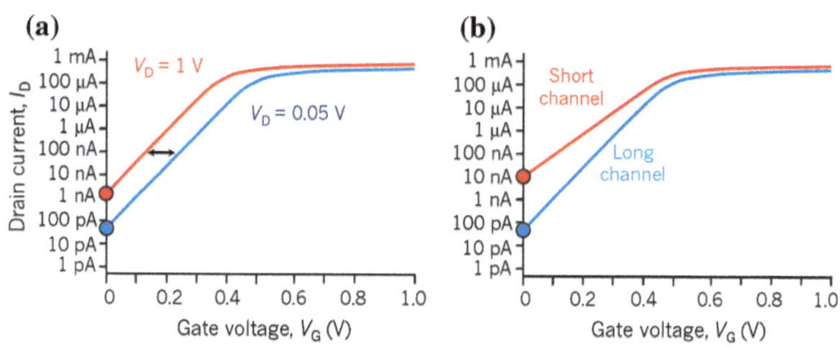

Fig. 1.4 Schematic diagram of transfer characteristics of devices in which SCE occurs. The blue curve shows the relation between the source-drain current I_{DS} and the gate voltage V_G at low source-drain bias ($V_D = 50$ mV); the red curve shows the relation between the source-drain current I_{DS} and the gate voltage V_G at high source-drain bias ($V_D = 1$ V). **a** Drift of threshold voltage caused by DIBL effect; **b** Increase of SS when the channel of devices becomes shorter. Both effects make the OFF-state current of the device larger. Reprinted from Ref. [1], with kind permission from Springer Nature

Generally speaking, the SCE is caused by an extended depletion layer near the drain electrode into the channel region when the device is pinched off [1]. Since the channel length of the device is small, the extension effect is not negligible, so that the effective gate length of the device is greatly reduced and the gate cannot control the conductivity of channel efficiently. This effect is more pronounced in the case where the drain voltage of the device increases. Consequently the potential distribution in channel is not only controlled by the gate voltage, but also depends on the channel length and the drain voltage. To reduce the SCE, it is necessary to enhance the gate tunability. Researchers use a natural length (λ) to characterize the length of the electric field lines that extends to the interior of the channel at the ends of the source and drain. As shown in Fig. 1.6, device performance is free from SCE when the channel length of the device is 4–6 times or more of λ. The length of λ differs in different device structures and is an indirect quantitative representation of the tunability of gates. Compared to the traditional planar structure as shown in Fig. 1.5a, the multi-gate structure increases the electrostatic control ability of the gate by enhancing the coupling between the gate electrode and the device channel. Taking the Si nanowire MOSFET with a square cross section as an example [1, 15], the λ of a conventional single-gate device can be simply simulated as $\lambda_1 = \sqrt{\frac{\varepsilon_{Si}}{\varepsilon_{ox}} t_{Si} t_{ox}}$; if a second gate is added to the lower part of the nanowire to form a dual-gate device, the λ of device is reduced to $\lambda_2 = \sqrt{\frac{\varepsilon_{Si}}{2\varepsilon_{ox}} t_{Si} t_{ox}}$; In the case of a gate-all-around structure, the λ will decrease to $\lambda_4 = \sqrt{\frac{\varepsilon_{Si}}{4\varepsilon_{ox}} t_{Si} t_{ox}}$. Here, ε_{Si} is the dielectric constant of Si, ε_{ox} is the dielectric constant of gate dielectric, t_{Si} is the side length of the square cross section of Si nanowires, and t_{ox} is the thickness of gate dielectrics. It can be seen that the most efficient gate structure of FET is the gate-all-around structure [16], as shown

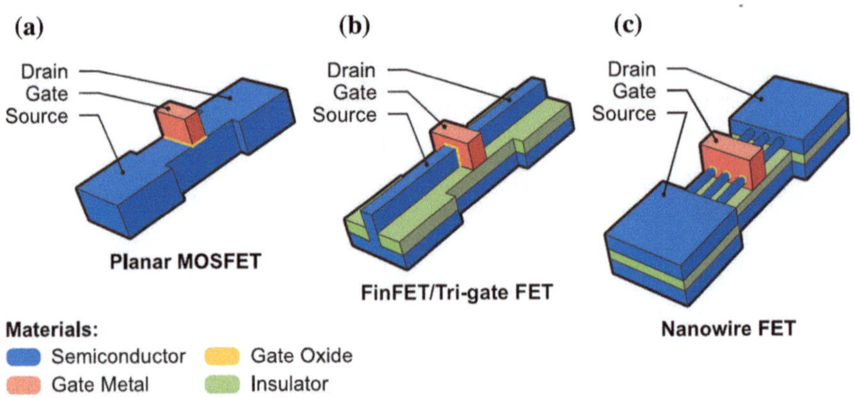

Fig. 1.5 The structure of FETs is developed from the leftmost planar gate structure to the middle triple gate structure and the rightmost gate-all-around structure, in order to improve the electrostatic control capability of gate. Reprinted from Ref. [5], with kind permission from Cambridge University Press

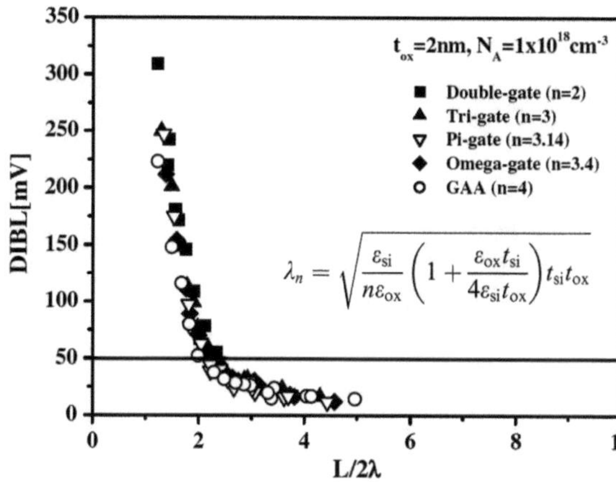

Fig. 1.6 The Scatter plot shows that the DIBL effect of a Si nanowire MOSFET having square cross section changes with L/λ_N. In the figure, the SCE begins to appear when L/λ_N is less than 6, and increases significantly when L/λ_N is less than 4. Reprinted from Ref. [15], Copyright 2007 with permission from Elsevier

in Fig. 1.6. Therefore, in terms of further scaling down, the gate-all-around structure is more favorable to fin- or tri-gate structures which have been already utilized in the most advanced commercial Si chips [1].

Nanowires are ideal for fabricating gate-all-around devices due to their cylindrical structure. Combined with its high electron mobility, InAs nanowires possess great potential in applications of high-performance electronic devices. In addition, since the semiconductor nanowires are only a few nanometers to a hundred nanometers in the diameter, the strain induced by lattice mismatch can be better released compared to bulk materials. It is more flexible to form axial or radial nanowire heterostructures. As shown in Fig. 1.7 [14], $In_{0.7}Ga_{0.3}As$ nanowires have a similar lattice constant as InAs, and they can be directly epitaxially grown on the Si substrate to form a radial heterojunction structure. Several other semiconductor materials such as InP and InAlAs can also be epitaxially grown in the axial direction of the InGaAs nanowires. This in turn has broadened the design for the research and development of new quantum devices or optoelectronic devices, e.g. TFETs.

1.2 Crystal Structure of InAs Nanowires

Bulk InAs has face-centered cubic zinc blende (ZB) structure (as shown in Fig. 1.8a). In low-dimensional InAs nanowires, hexagonal wurtzite (WZ) structure as shown in Fig. 1.8b also exists [17, 18]. These two crystal phases, ZB and WZ, of InAs material have very similar formation energy. As reported by theoretical calculation, the difference of formation energy between them are just on the order of 10 meV/atom [19]. Therefore, when the proportion of surface energy in total energy of material system increases due to the continuously scaling down of material, WZ structure which has a lower surface energy will replace ZB structure and become the stable phase with lower energy [20]. Glas et al. calculated that the InAs nanowires along

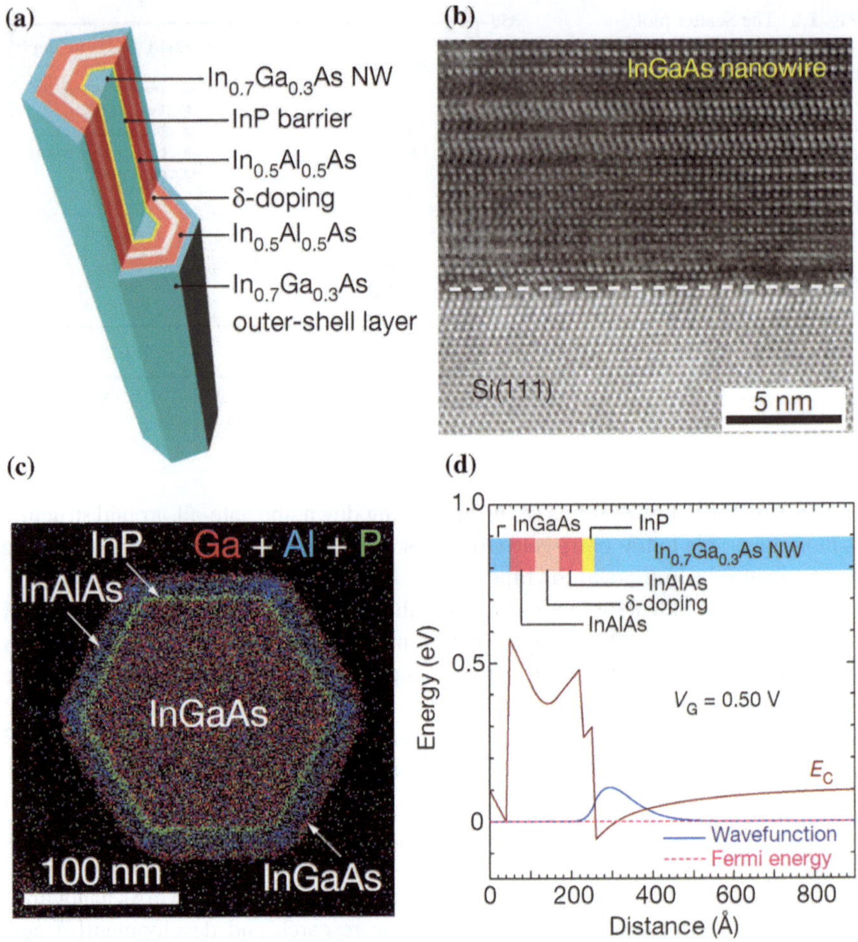

Fig. 1.7 $In_{0.7}Ga_{0.3}As$ nanowires having a similar lattice index with InAs are epitaxially grown in the axial direction by precisely controlling gas source and temperature, in order to form a quantum well structure that has similar structure with planar two-dimensional electron gas. **a** Schematic diagram of material composition of $In_{0.7}Ga_{0.3}As$ nanowires; **b** TEM image of the interface between $In_{0.7}Ga_{0.3}As$ nanowire and Si substrate; **c** Two-dimensional scanning of composition of $In_{0.7}Ga_{0.3}As$ nanowires at cross section; **d** Schematic diagram of band structure of $In_{0.7}Ga_{0.3}As$ nanowires. Reprinted from Ref. [14], with kind permission from Springer Nature

<111> direction present a transition from ZB to WZ phase at a critical diameter between 10 and 20 nm [20]. When the diameter of InAs nanowires is just slightly larger than this critical point, its arrangement of atomic layer along the axial direction is often mixed between ZB and WZ phase at nanometer scale, [21, 22] leading to many stacking faults and mixed phases perpendicular to the nanowires' axial direction.

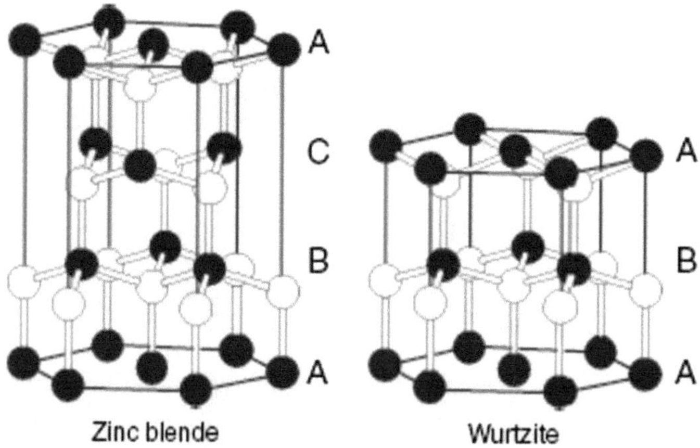

Fig. 1.8 Two crystal structures of InAs nanowires. The left picture shows the face-centered cubic ZB structure, which is the ABCABC... stacked structure; the right picture shows the hexagonal WZ structure, which is the ABAB ... stacked structure. Reprinted with permission from Ref. [23]. © 2008 by American Physical Society

Pan et al. have reported that when molecular beam epitaxy (MBE) is used to grow <111> oriented InAs nanowires with Ag particles as catalyst, as shown in Fig. 1.9, nanowires with diameter less than 48 nm are pure WZ phase; when the diameter of nanowires falls in the range from 48 to 72 nm, a large number of stacking faults appear along the growth direction of nanowires, making the entire nanowire a mixed phase structure of WZ and ZB; when the diameter continues to increase, the structure changes to pure ZB phase. By controlling the conditions of growth, such as the temperature and the amount of gas source, as shown in Fig. 1.10, Kimberly et al. have reported a method that use metal-organic chemical vapor deposition (MOCVD) to control the alternate growth of ZB and WZ structures in large-diameter InAs nanowires [17] [23].

The crystal structure of InAs nanowires is related to their growth method [24], growth conditions [25, 26], growth direction [27, 28], and diameters [22, 26]. In previous works, InAs nanowires grown along the <0001> and <0-110> directions have been reported to show WZ phase [22, 26], while nanowires grown along other directions have ZB phase [22, 26]. Currently, InAs nanowires grown along the direction of ZB <111> and WZ <0001> are most controllable. Nanowires grown along ZB <111> or WZ<0001> direction may have different structures under different growth conditions: when the growth temperature is relatively low or the diameter is relatively large, the nanowires tend to have ZB structure; on the contrary, when the growth temperature is relatively high or the diameter of is relatively small, they are more likely to become WZ phase; under the intermediate condition, the nanowires form a mixed phase structure [22].

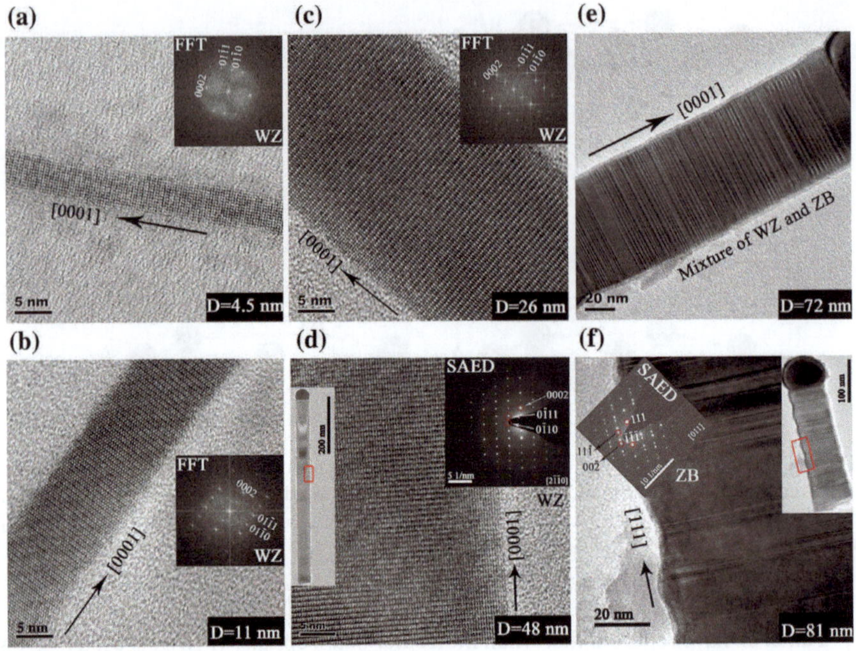

Fig. 1.9 High-resolution TEM images and their Fast Fourier Transform Algorithm (FFT) of different diameter InAs nanowires catalyzed by Ag particles. Nanowires are grown on Si (111) substrates by MBE systems. Their growth direction is <111>. The diameters of InAs nanowires in **a**–**f** are 4.5, 11, 26, 48, 72 and 81 nm, respectively, and their crystal structure first changes from pure WZ phase to mixed phase, and then gradually transforms into pure ZB phase. Reprinted with the permission from Ref. [22]. Copyright 2014 American Chemical Society

Regardless of ZB or WZ structure, as shown in Fig. 1.11, InAs nanowires are all semiconductor materials with direct band gaps. However, due to the different atomic arrangement of these two kinds of crystal structures, the WZ and ZB structure of InAs nanowires have different band structures. Many theoretical and experimental studies have confirmed that the bandgap of WZ structure InAs material is tens of meV to hundreds of meV larger than that of ZB structure [29–32]. However, several issues have not been studied in detail such as whether Fermi level pinning exists on the surface of WZ phased InAs or its interface with metal, and if the position of pinned fermi level is at the bottom of the conduction band, etc. [33–35]. It is still controversial whether the properties of Fermi pinning on the surface of ZB phase InAs material changes with the size reduction [36, 37].

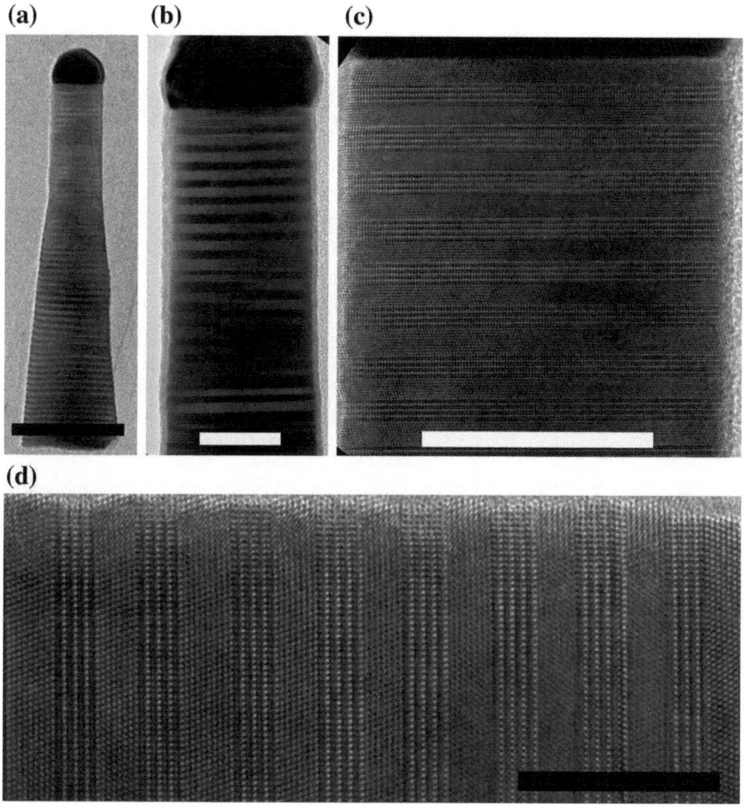

Fig. 1.10 TEM characterization **a**-**d** of InAs nanowire having alternating WZ and ZB phase grown by MOCVD [17]. The scale bar in **b** and **c** are both 20 nm. Reprinted with the permission from Ref. [17]. Copyright 2010 American Chemical Society

1.3 Development Status of InAs Nanowire Electronic Devices

The study of nanowires dated back to 1964 [38], but the study of InAs nanowires did not began until around forty years later [39]. In the past ten years, scientific research based on InAs nanowires has continuously made new progress in the areas of electronic devices [5, 40–42], novel quantum devices [43–46], infrared detection [47–49] and solar cells [50]. In the aspect of InAs nanowire-based electronic device, high-performance MOSFETs based on InAs nanowires and TFETs in combination with other semiconductors have become the subject of current interest. In the following, we will introduce the performance of these two kinds of electronic devices based on single InAs nanowire or nanowire arrays, and then summarize the research progress in recent years.

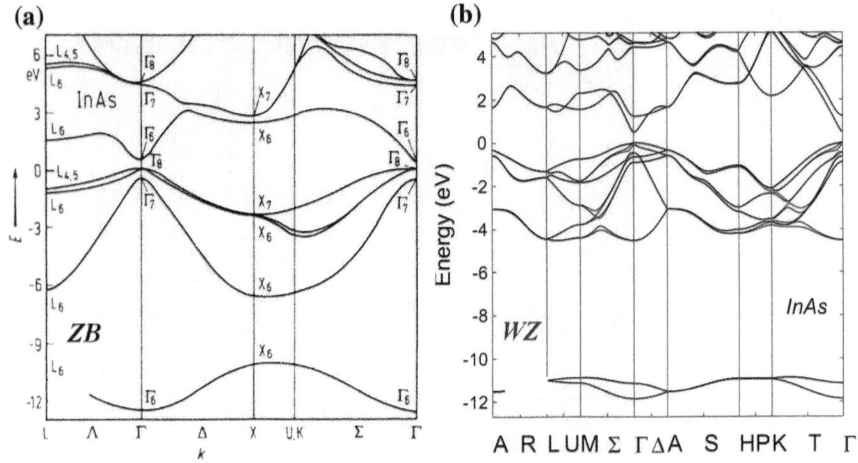

Fig. 1.11 **a** Bandstructure of ZB phase InAs bulk material; **b** Bandstructure of WZ phase InAs material obtained by theoretical calculation. Reprinted with permission from Ref. [32]. Copyright 2010 by American Physical Society

1.3.1 InAs Nanowire MOSFETs

As shown in Table 1.2, the ON-state performance of MOSFETs based on a single InAs nanowire has reached comparable level of HEMTs and Si-based FinFETs. Dey et al. have grown dumbbell-like InAs nanowires by controlling growth conditions and doping environment [41]. For these nanowires, the channel part have diameter of 15 nm, and in order to reduce the scattering of electrons by impurities, the channel part is pure WZ phase without intentional doping; while the contact portion uses highly doped ZB phase InAs nanowires having diameter of about 40 nm to reduce contact resistance between nanowires and metal electrode. Based on these dumbbell-shape nanowires, Dey et al. obtained the I_{on} of 0.6 A/mm and maximum transconductance (g_m) of about 1.23 S/mm by facile fabrication process.

Table 1.2 Comparison between ON-state properties of FETs having various dimensions and structures. The data in this table are all extracted under the condition of $V_{DS} = 0.5$ V (or similar to 0.5 V). The I_{on} and g_m of nanowire devices are normalized by perimeter of nanowires [41]

D (nm)	L_G (nm)	I_{on} (A/mm)	$g_{m, max}$ (S/mm)	SS (mV/dec)	References	Tech.
15	100	0.6	1.23	140	[41]	InAs NW
10	30	0.7	1.9	80	[8]	InAs HEMT
30	250	0.12	0.56	120	[53]	Si FinFET
10	75	0.55	1.75	95	[54]	InAs QWFET
25	170	0.4	0.8	260	[100]	Radial InAs NW
13	230	0.9	1.72	180	[113]	InAs films

In addition to exploring high-performance FETs based on single InAs nanowire, researchers have also conducted researches on MOSFETs devices based on InAs nanowire arrays [40, 55–62] and CMOS devices integrated with other materials [51, 52, 63], which may pave the way for practical applications of InAs nanowire high-performance devices.

Since InAs nanowires can be uprightly grown on III–V [64, 65] or Si substrates [66, 67] (typically (111) faces), they are well suitable for vertical gate-all-around devices. Persson et al. fabricated vertical gate-all-around devices based on a single InAs nanowire and showed I_{on} of 1.34 A/mm and a g_m of 1.37 S/mm which are similar to those of planar InAs nanowire-based MOSFETs [60]. However, because the electric current of a single nanowire is too small, they cannot be directly used in the circuit, so Johansson et al. fabricated gate-all-around RF devices based on InAs nanowire arrays taking advantage of the array growth [42]. In the growth procedure, metal particles [65] or SiO_2 patterns [14] are used as templates. The g_m of the nanowire array gate-all-around devices can also reach 730 mS/mm. By using finger contacts instead of the conventional pad contacts, the parasitic capacitance is significantly reduced. As a result, the cutoff frequency (f_T) of nanowire array-based devices can be improved to 103 GHz, and the maximum oscillation frequency (f_{max}) is 155 GHz. These values are the highest in current RF devices based on InAs nanowires. Moreover, Persson et al. of the same research group have verified the feasibility of their InAs nanowire vertical gate-all-around devices in the circuit, by connecting three array gate-all-around RF devices with two resistors to form a mixer circuit [68].

Although the InAs material excels in electrical properties and is very suitable for fabrication of n-type high-frequency devices, it is not suitable as the channel material for p-type devices due to its poor transport properties of holes [69]. Therefore, in addition to merely using InAs nanowires to fabricate electronic devices, researchers also combined InAs nanowires with other semiconductor nanowires to form CMOS devices on chips [51, 52, 63]. As shown in Figs. 1.12 and 1.13, single nanowire-based planar and nanowire array-based vertical CMOS devices both employed InAs nanowires as the channel material of n-type MOSFET, while GaSb nanowires with high hole mobility are used for p-type FET [51, 52]. Although both these planar and vertical nanowire-based CMOS devices have been shown to function as the inverter to conduct the basic operation, the performance of these CMOS devices is still far behind the existing microelectronics industry, mainly because the n-type and p-type MOSFETs in the CMOS are not matched and they also both have defects in the OFF-state characteristics.

1.3.2 InAs Nanowire TFETs

Three-dimensional FETs using novel materials have great potential in improving the ON-state performance of devices, but due to the limitation of the diffusion current in the FETs, the theoretical limit of SS at room temperature is 60 mV/decade. This is another large factor that limits the scaling down of V_{DD} [1]. To overcome this limit,

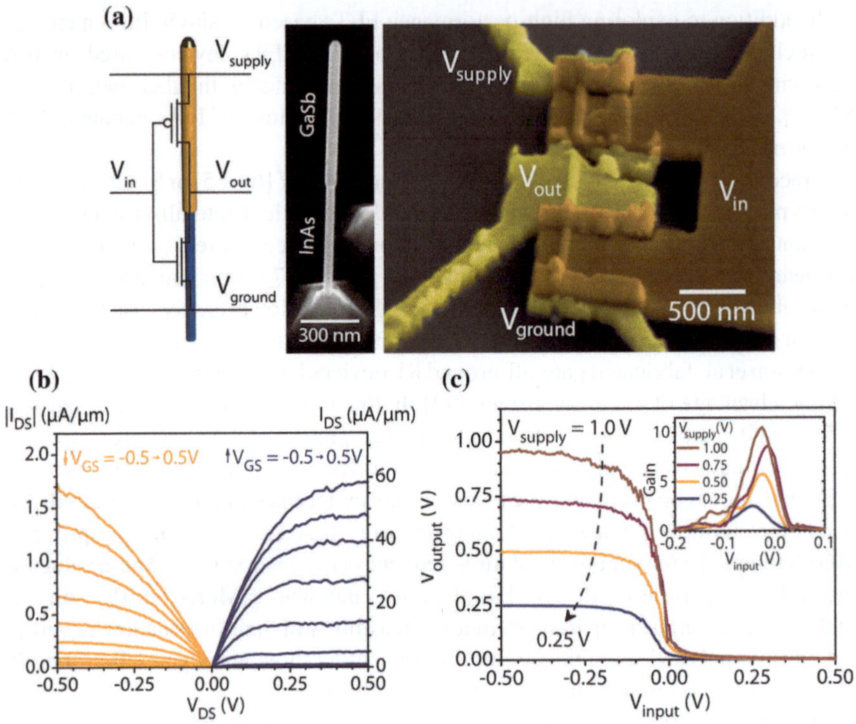

Fig. 1.12 **a** The nanowire in the diagrams can formed by the following steps: first, epitaxially grow InAs nanowires on the substrate, and then epitaxially grow GaSb nanowires on the InAs nanowires. The InAs nanowires are used as the channel material of n-type MOSFET, and the GaAs nanowires with high hole mobility are used as the channel material of p-type MOSFET. The SEM image of nanowire-based top-gated device is taken at a 30° tilted angle. **b** Output curves of InAs n-FET (blue) and GaSb p-FET (orange). As can be seen from the figure, the electrical performance of the n-FET is much better than that of the p-FET, and the performance of these two FETs do not match. Moreover, the threshold voltage of both n-FET and p-FET cannot meet the requirements for low-power CMOS; **c** Voltage transfer characteristic of CMOS at different operating voltages. This feature of the CMOS devices also cannot meet the demands of current circuit design. Reprinted with the permission from Ref. [51]. Copyright 2012 American Chemical Society

researchers have proposed to replace MOSFETs with TFETs as the basic elements of logic circuit [70]. The TFET controls the current of device by tunneling mechanism, which avoids the influence of diffusion current on the OFF-state of devices and thus enables a smaller SS (SS < 60 mV/decade) [71]. However, since the I_{DS} of TFETs at the ON-state is mainly contributed by band to band tunneling (BTBT), I_{on} of TFETs is typically smaller than that of MOSFET. In order to increase the I_{on} of TFETs, using materials with low carrier effective mass becomes one of the choices for optimizing the performance of TFETs. Due to the small electron effective mass, InAs material is very suitable for high performance TFETs.

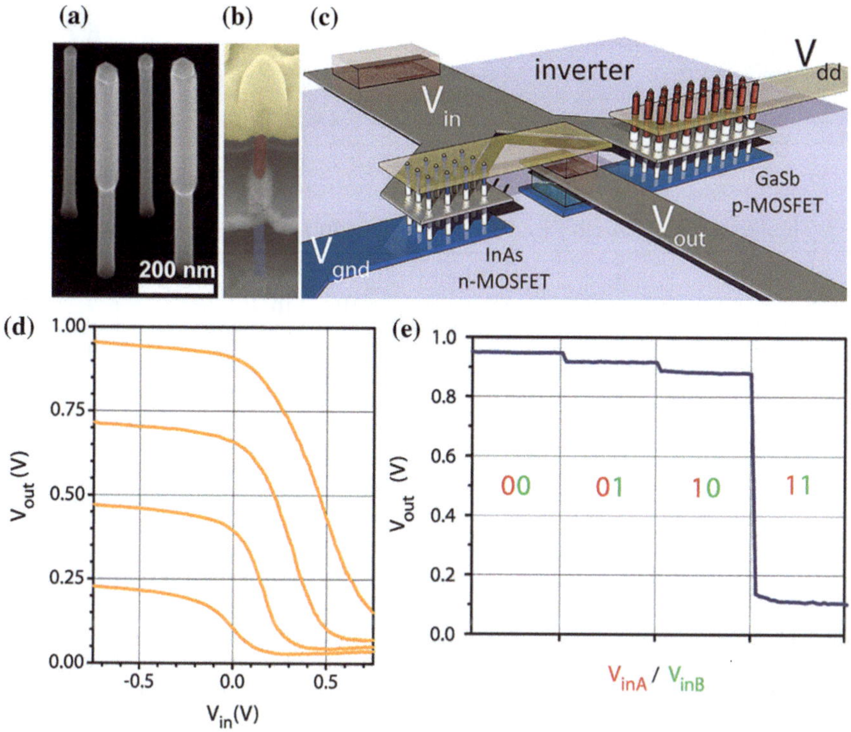

Fig. 1.13 **a** SEM image of vertically grown InAs nanowires and InAs-GaSb nanowires. **b** false-color SEM image of the cross section of vertical gate-all-around InAs nanowire devices. **c** Schematic diagram of a CMOS composed of InAs nanowire array and InAs–GaSb nanowire array gate-all-around transistors. The InAs nanowire array-based transistors act as the n-FETs, and the InAs–GaSb nanowire array-based transistors are p-FETs. **d** Voltage transfer characteristics of CMOS at different operating voltages. **e** Operating characteristic curve of NAND composed of vertical gate-all-around CMOS under the condition of operating voltage $V_{DD} = 1$ V Reprinted with the permission from Ref. [52]. Copyright 2015 American Chemical Society.

InAs nanowires together with Si [61, 72], Ge [40], GaSb [73, 74], InSb [46] and other materials can form a tunnel junction as basis for TFETs. Figure 1.14 shows TFETs based on InAs-GaSb nanowire heterojunctions and some of these devices achieve I_{on} of 10^4 kA/cm^2 at $V_{DS} = 50$ mV [73]. However, TFETs with high ON-state current are often compromised in OFF-state performance [73, 74]. Figure 1.15 shows the I_{on}–I_{off} relation for TFET devices as of 2014 [5]. As shown in the figure, the existing TFETs basically satisfy the law that I_{off} increases monotonically with I_{on}. Therefore, for the current InAs nanowire TFETs with relatively good ON-state performance, it is necessary to further study the topic of reducing the OFF-state current of TFETs.

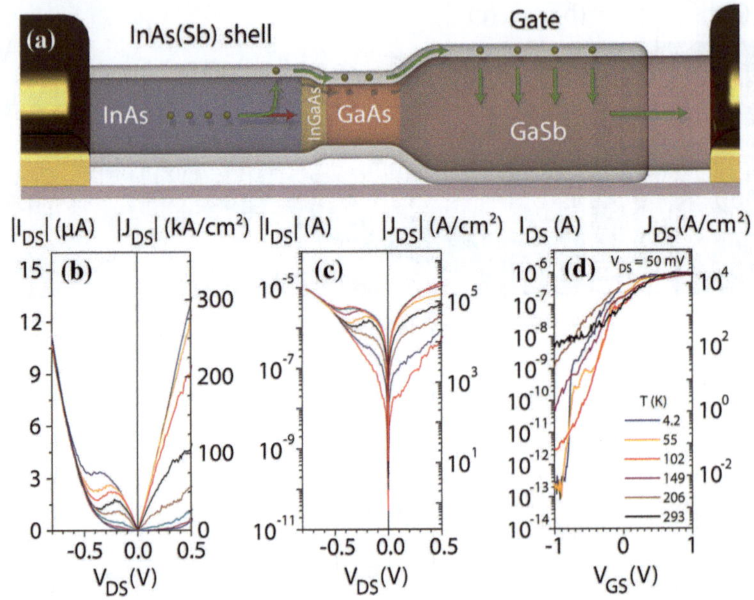

Fig. 1.14 **a** Schematic diagram of current conduction in a radial TFET. In the figure, the current conducted from the core of InAs nanowire is blocked by the barrier formed at the heterojunction between InAs and GaSb, so that the current has to radially flow out to the outer InAsSb shell. After passing through the heterojunction, the current then flows back to the core GaSb through the radial barrier of InAsSb–GaSb, **b** and **c** plots of the I_{DS}–V_{DS} output for this TFET in linear and exponential coordinates, **d** is the transfer characteristics of device at different temperatures. The current transport mechanism changes when the temperature is lower than 102 K. In **d**, the device is at a reverse bias state, the InAs terminal is connected to a drain voltage of 50 mV, and the GaSb terminal is grounded. Reprinted with the permission from Ref. [73]. Copyright 2013 American Chemical Society

1.4 Several Problems of InAs Nanowire-Based Electronic Devices

The potential of InAs nanowire for n-type high-performance electronic devices are widely recognized. But there are still several questions or problems remaining unclear or unsolved, such as the integration with existing Si processes, instable performance of devices, controlled growth of high quality materials, high OFF-state currents, large density of interface states, impacts of reduced diameter on device performance etc.

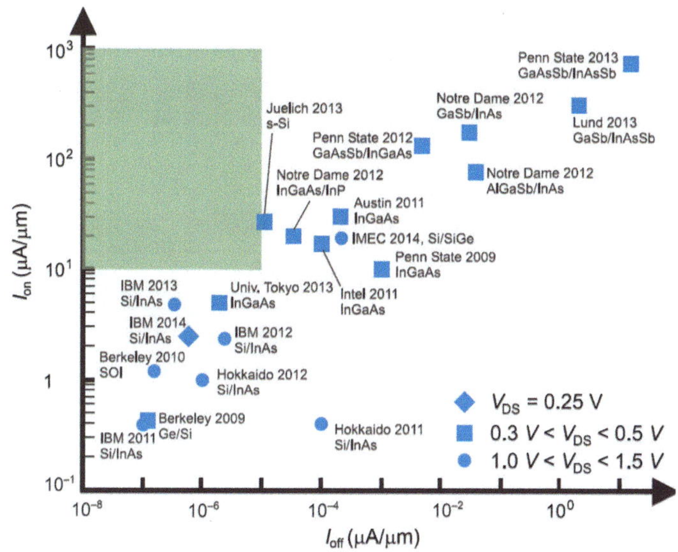

Fig. 1.15 The I_{on}–I_{off} relation of TFET devices as of 2014. In the figure, the existing TFETs basically satisfy the relation that the I_{off} increases monotonically with I_{on}. Reprinted from Ref. [5], with kind permission from Cambridge University Press

1.4.1 High OFF-State Current of Devices

As discussed in previous section, the problem of high OFF-state current exists, in both InAs nanowire MOSFETs and TFETs [41, 42, 73, 75]. This is closely related to the characteristics of intrinsic band structure of InAs. As one of the high electron mobility, narrow bandgap semiconductors, InAs has a bulk band gap of only 0.36 eV. As a result of this small bandgap, it is difficult to reach the industry's required 100 nA/μm for the OFF-state current, even in long-channel MOSFETs [76, 77]. And it leads to the problem of large static power consumption of devices [78].

In short channel and large source-drain bias conditions, the carriers of narrow bandgap semiconductor device can obtain high enough energy under the acceleration of a strong transverse electric field to collide with the atoms of semiconductor material in the channel. Thus, the electrons can be excited to conduction band and electron-hole pairs are formed. If the generated electrons and holes acquire sufficiently high energy under the acceleration, then they will continue to generate new electron-hole pairs by further collisions, thereby producing a sharp increase in the OFF-state current. This process is called impact ionization [79]. For the MOSFET, since the voltage potential mainly drops across the device contact at the OFF-state, the impact ionization will also be the most intense at the contacts, which renders weakened potential barrier. The smaller the semiconductor band gap, the lower the energy required for the generation of electron-hole pairs. To circumvent this, it is desired to increase the bandgap of material.

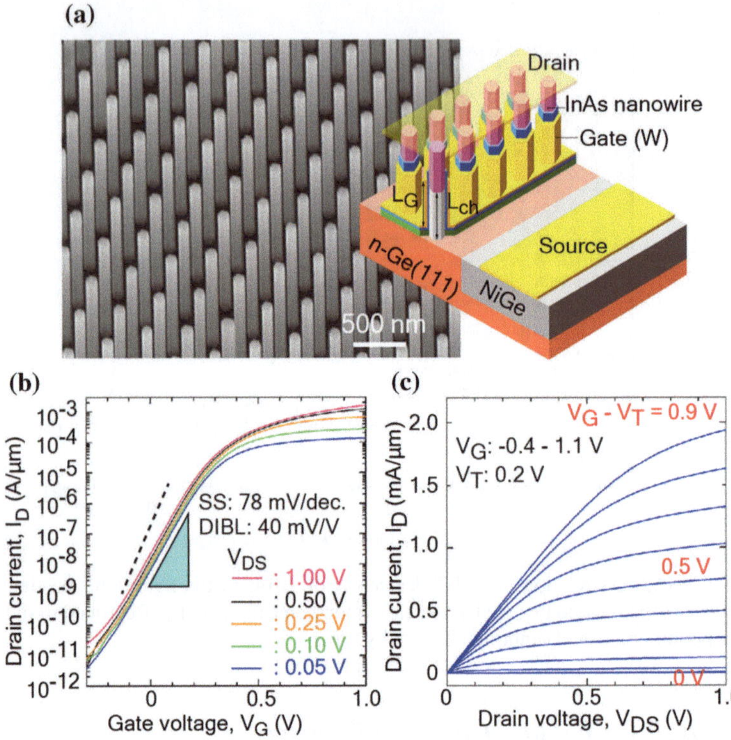

Fig. 1.16 **a** The left diagram is the SEM image of InAs nanowire array, and the right one is the schematic diagram of the vertical gate-all-around MOSFET based on the nanowire array. **b** Transfer characteristics of the device; **c** Output characteristics of the device. Reprinted with the permission from Ref. [40]. Copyright 2015 American Chemical Society

One method used by researchers is to design a heterojunction at the contact between the semiconductors that have relatively large bandgap and InAs material. As shown in Fig. 1.16, Tomioka et al. have effectively avoided the effects of impact ionizations and band-to-band tunneling by making a heterojunction between InAs nanowires and n-type Ge (111) substrate [40]. At the same time, by heavily doping the n-type Ge substrate, the contact resistance is reduced. Although the method shown in Fig. 1.16 can effectively reduce the OFF-state current without reducing the ON-state current of the devices, the process of fabrication is rather complicated, and Ge substrate is incompatible with the current Si process. In addition, the InAs nanowires grown directly on Ge in this work are not among the highest quality because of their large number of stacking faults.

In Sect. 1.2, we have mentioned that the WZ phased InAs nanowires have a larger energy gap than that of ZB phase [29], and boths phase can be alternately controlled in the same nanowire by controlling growth conditions [17]. If the energy band of semiconductor material can be designed by using different crystal structures of the same material, the requirements for material growth will be greatly eased, meanwhile the device fabrication process will be more flexible. In addition, because of the large Bohr radius of InAs material, the bandgap of nanowires might increase during scaling down because of the quantum confinement effects [80], which is also beneficial for reducing the OFF-state current.

1.4.2 Interface State Between InAs Nanowires and Gate Dielectric

The high-quality interface between SiO_2 and Si is one of the important reasons why Si materials became the industrial chip materials in the past decades. However, as the size of device has reached nanometer level in recent years, the thickness of SiO_2 gradually decreases and the tunneling current from the gate significantly increases; meanwhile, the strong electrical field in gate dielectrics is close to breakdown. To solve this problem, in the latest generations of Si process nodes, a hafnium-based high-k dielectric layer was developed to replace the traditional SiO_2 dielectric layer. By this method, excessive tunneling current caused by thin gate dielectrics can be sufficiently reduced. At present, it has become a standard to use high quality high-k gate dielectrics in high performance electronic devices.

High-k dielectrics include HfO_2, ZrO_2, Al_2O_3, Y_2O_3 etc., and the deposition methods include atomic layer deposition (ALD) [81], plasma enhanced chemical vapor deposition (PECVD) [82], metal film thermal oxidation [83] etc. By using ALD to deposit the gate dielectric layer for InAs electronic device, a dense and less defective gate dielectric film can be obtained, and also the surface is cleaned and passivated during the growth process by selecting specific kind of growth precursor, so that the interface between InAs material and high-k dielectrics can be optimized [84]. At present, high-k materials used for the gate dielectrics of InAs nanowire electronic devices mainly include: HfO_2, ZrO_2, Al_2O_3 etc. Although the interface between InAs materials and high-k dielectrics have been improved over the past years, there are still problems to be discussed next (Fig. 1.17).

Wheeler et al. studied the interface between HfO_2 and InAs and found that for the metal—HfO_2-InAs (Au/Ti/HfO2/InAs) plate capacitor structure, the density of interface states can be controlled at 3×10^{12} cm^{-2}, after several processes of optimization such as etching the native oxide layer of InAs, annealing under the N_2 atmosphere etc. [86]. According to recent works by Takei et al., the density of interface states between ZrO_2 and InAs can be decreased down to 1.1×10^{12} cm^{-2} when a low temperature (170 °C) annealing process under the atmosphere of forming gas (generally H_2 and O_2 combined gas) is used [85]. This is one of the minimal interface state density achieved, but this density is still much higher than that at interface of Si/SiO_2, which is on the order of 10^{10} cm^{-2}. In addition, in order to improve the stability of gate dielectrics, researchers also proposed to replace the monolayer binary dielectrics with a combination of multiple layered gate dielectric [87], intrin-

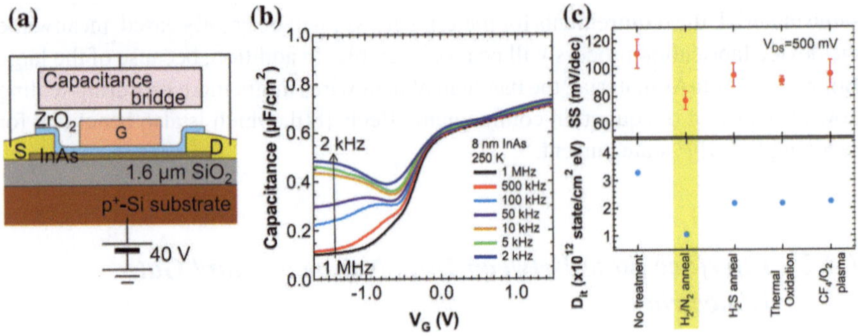

Fig. 1.17 **a** Schematic diagram of device structure and measuring apparatus for C–V and G/x–f measurements; **b** C–V characteristics of InAs film-oxide-insulator (semiconductor film-oxide-insulator, XOI) FET structure at temperature T of 250 K. This sample was annealed in forming gas after ALD process. **c** The average subthreshold swing and density of interface states of XOI FET devices after different interface processing methods. Reprinted with permission from Ref. [85]. Copyright 2013, American Institute of Physics

sic thermal oxide layer transition interface [88] and multicomponent oxides [40, 89]. However, all these methods have not yield the interface quality comparable to state-of-the-art Si process. In addition, the interface between WZ phase InAs material and gate dielectrics is less understood.

The surface (or interface) properties plays an important role in the properties of band structure of InAs NWs [29, 30, 90]. When the surface of InAs nanowires is totally unpassivated, the defects and dangling bonds on the surface lead to a decrease of bandgap [30, 90]. Because the Fermi level is pinned at the bottom of conduction band [33, 35] on the unpassivated InAs surfaces, the threshold voltages of most InAs nanowire MOSFETs are mostly negative or around 0 V [26, 41, 51, 91, 92], which leads to the problem to the input voltage of CMOS (as shown in Figs. 1.12 and 1.13) [51, 52]. If the Fermi pinning or interfaces traps are significant on the surface, the electrostatic coupling between the gate and the nanowire will also be affected [79]. Therefore, the quality of the interface between InAs and dielectrics has a major influence on the performance of InAs nanowire electronics [93], particularly for TFETs. It is one of the most important issues to improve the quality of interface between gate dielectrics and InAs nanowires.

1.4.3 *Relation Between Structure of InAs Nanowires and Their Device Performance*

Although the introduction of a high-k gate dielectrics allows the gate dielectric layer to enhance the electrostatic coupling while maintaining its physical thickness, however, for nanowire FETs, the performance of devices still degrades due to the SCEs. To prevent the degradation, the diameter of nanowires should also be reduced accordingly when the channel length decreases. For InAs nanowires the cross section is circular [36] or a hexagonal [94], the natural length λ of the gate-all-around FET is: [16].

$$\lambda_{wrap} = \sqrt{\frac{2\varepsilon_{InAs} D_{InAs}^2 \ln\left(1 + \frac{2t_{ox}}{D_{InAs}}\right) + \varepsilon_{ox} D_{InAs}^2}{16\varepsilon_{ox}}} \tag{1.2}$$

Here ε_{InAs} is the dielectric constant of InAs nanowires, D_{InAs} is the diameter of InAs nanowires, t_{ox} and ε_{ox} are the thickness and dielectric constant of the gate dielectrics, respectively. According to Dey's work, to obtain an effective channel length below 10 nm while effectively avoiding the SCEs, the diameter of InAs nanowires at the channel need to be less than 10 nm [95].

A few studies reported on InAs nanowire FETs with diameters less than 10 nm [36, 41, 91]. Razavieh et al. reported back-gated devices based on 10 nm InAs nanowires [36]. The work of Razavieh et al. pointed out that the bandgap of InAs nanowires with diameter of 10 nm is about 120 meV larger than that of 30 nm [96, 97]. This is because of enhanced quantum confinement phenomena [96, 97] at small diameters. Razavieh et al. also mentioned that, unlike InAs bulk material, nanowires with diameter of 10–30 nm form a Schottky barrier of about 110 meV with the metal Ni. The Schottky barrier reduces the ON-state current of the electronic devices and thus is detrimental to the performance of devices. Yet other groups reported contradictory results [37, 98], so further work is needed to verify this conclusion.

Furthermore, due to the quantum confinement effects and severe surface scattering [80], the electron mobility decreases as the diameter of nanowires. A.C. Ford et al. reported that the electron field-effect mobility (μ_{FE}) in long-channel FETs based on InAs nanowires with diameter larger than 15 nm is linear to the diameter, as shown in Fig. 1.18. Therefore, the electrical properties of ultrathin InAs nanowires are very sensitive to the diameter.

As nanowires become smaller, the crystal structure of nanowires grown along <0001> direction tends to be in the WZ phase [22]. According to previous works, the crystal phase (WZ phase or ZB phase) is also one of the important factors that affect the electrical transport properties of InAs nanowires. And its influence on the band structure and device performance have been extensively studied [17, 26, 29, 31, 32, 99, 100]. A. Deyeh et al. reported that pure ZB phase InAs nanowires grown along the <011> direction show a lower ON–OFF ratio and a similar μ_{FE} compared with the WZ phase <0001> grown nanowires containing a large number of stacking faults [99]. However, Ullah et al. pointed out in their work that the μ_{FE} of pure WZ phase InAs nanowires grown along the <0001> direction is higher than that of pure ZB phase InAs nanowires grown along the <111> direction [100]. Ullah et al. also pointed out that A. Deyeh et al. inappropriately attributed the higher ON–OFF ratio of WZ phase InAs nanowires to the polarization charge and barrier caused by stacking faults in nanowires [100]. However, theoretical calculations show that WZ phased InAs nanowire has a larger effective mass than the ZB phase nanowires [32]. In ideal

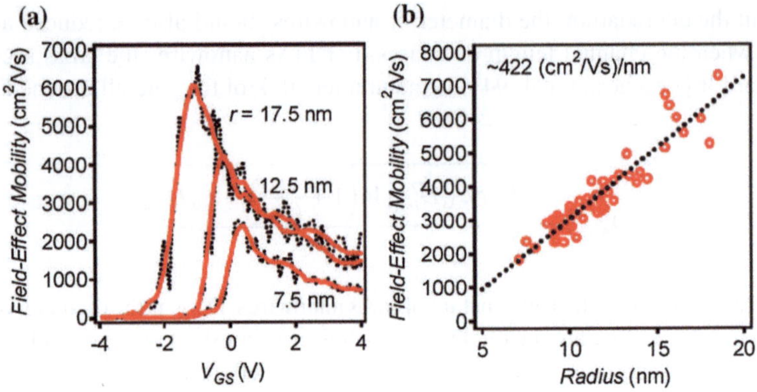

Fig. 1.18 **a** Relation between the gate voltage and field-effect electron mobility of InAs nanowire FETs with different radius. In the figure, the InAs nanowires have radius of 7.5, 12.5 and 17.5 nm, respectively. **b** Maximum field-effect electron mobility of InAs nanowire FETs changes with radius of nanowires at room temperature. Reprinted with the permission from Ref. [91]. Copyright 2009 American Chemical Society

conditions, the electron mobility of the WZ phase InAs nanowires is lower than that of the ZB phase nanowires [32]. Since many conclusions of publicated works are contrary to each other, the effects of crystal structures on the electrical transport properties of InAs nanowires (such as electron mobility and electrical resistance) are still unclear [17, 26, 29, 31, 32, 99, 100].

According to theoretical calculations, even for InAs nanowires of the same crystal structure grown along different crystal orientations, different band structures may appear [30, 101, 102]. For example, although the <001>, <011>, and <111> directions are all low-index growth directions, small-diameter InAs nanowires may have smaller effective mass, higher mobility, and weaker quantum confinement effects in some particular directions [30, 101, 102]. Moreover, both recent theoretical calculations [103] and experimental results [104, 105] have confirmed that InAs nanowires with different crystal orientations show different piezoelectric and piezoresistive phenomena, due to the different order and polarity of In and As atoms in different crystal phases or orientations. This points to using nanowires grown along specific directions to fabricate devices to increase the operation speed and the stability of the devices. In terms of key issues such as which crystal direction to be used, theoretical calculations showed large discrepancy [30, 101, 102]. It is difficult to accurately predict the band structure of nanowires grown along different directions by theoretical calculations if the key input parameters for calculation cannot be accurately obtained from experimental data. On the other hand, the relation between crystal orientation and electrical transport properties of nanowires is still unclear. Although some groups have been able to grow pure phase ZB <001> [27] and ZB <121> [106] InAs nanowires, the electrical properties of various growth directions have not been characterized due to various reasons.

Stacking faults (and the resulting mixed phase structure) as a crystallographic disorder have been confirmed to severely reduce the carrier lifetime of III–V nanowires, and cause a considerable impact on the optical properties [107, 108] and electrical transport [99, 109] of nanowires. Many researchers reported on the effects of stacking faults (or mixed phases) on the electrical properties of InAs nanowires, most of which focused on electron mobility, conductivity and ON–OFF ratios of nanowires [26, 29, 99, 109]. Regarding the effect of stacking faults on ON–OFF ratio, Dayeh et al. proposed that the resistivity and ON–OFF ratios of InAs nanowires will increase due to the presence of polarized charges and barriers at the interface between ZB and WZ mixed-phase which help large-diameter InAs nanowires deplete the carriers therein [26, 99]. However, researchers have used the scanning tunneling microscopy (STM) to show that no polarization charge exists at the interface between the two phases. Therefore, the mechanism of influence of stacking fault on the ON–OFF ratio is unclear. At present, the question of how to avoid stacking faults in InAs nanowires with arbitrary diameter by controlling growth conditions remains unsolved.

Because several structural parameters such as diameter, crystal phase, growth direction and stacking faults often appear simultaneously, the influences of these factors are rarely separated explicitly in the existing experiments studying the properties of InAs nanowires. Therefore, when studying the effects of these factors on the electrical performance of InAs nanowires, how to avoid the interference from other factors is also one of the key points and difficulties in the study of electrical properties of InAs nanowires.

1.4.4 Dispersion of Device Performance

Due to the relatively high density of interface (or surface) states [73], large critical size of quantum confinement effects [76, 98], the mixed structure [26] of the InAs material etc., the devices based on InAs nanowires often show a very large dispersion on their ON-state or OFF-state performance. This large performance dispersion is very harmful to further integration of devices. Persson et al. report that, the normalized performance of a single InAs nanowire device is significantly better than that of the InAs nanowire array device [60]. Therefore, in order to obtain better performance, a good consistency of the performance of devices are important.

At present, a number of groups have made significant progress in the growth of high-quality InAs nanowires [22, 27, 28, 65, 110–112]. By using metal catalysts to guide the growth of nanowires, researchers have been able to grow pure phase InAs nanowires along <111> or <0001> directions at controlled positions of a III–V substrate (or III–V film) and the diameter of nanowires can be controlled by the size of catalyst particles [57, 65].

1.5 Topic Ideas and Chapter Arrangements

To briefly summarize, the study on InAs nanowire device is important for the application of novel device structures. Despite the many breakthroughs, it still faces a series of problems. In this thesis, researchs are selected on the following issues:

1. **Effect of scaling down on the electrical transport properties of InAs nanowires**. In this thesis, InAs nanowires with diameter less than 10 nm are studied, mainly due to the situation that the scaling down is very necessary to continue the "Moore's Law" and if the nanowires are taken as the channel material, their diameter should be reduced to nano-level in order to surpass the state of art of current Si process. However, in reported works of InAs nanowires, only a few research groups have successfully grown InAs nanowires having diameter less than 10 nm, and the studies on the performance of their devices are almost none due to the strict process conditions for fabricating devices based on ultrathin nanowires. In this thesis, ultrathin InAs nanowires with diameter less than 10 nm (minimum diameter up to 7 nm) grown by the team of Prof. Jianhua Zhao (Institute of Semiconductors, Chinese Academy of Sciences) are used to fabricate MOSFETs through the precisely controlled fabrication process, and then the performance of fabricated nanowire devices are systematically studied.
2. **Effect of crystal phase and orientation on the electrical transport properties of InAs nanowires**. Due to the large surface to volume ratio, nanomaterials often exhibit anisotropic characteristics. However, the impact of the crystal phase and orientation of InAs nanowires on device performance is still unclear. In this thesis, a novel method is developed to achieve accurate correspondence between structure and performance. By combining micro-nano processing technology, in situ nano-manipulation technology in scanning electron microscopy, and transmission electron microscopy characterization, the InAs nanowires fabricated into devices and electrically measured under variable temperatures are transferred to transmission electron microscopy for atomic-level characterization of their crystal structure. Using this method, the electrical transport properties of InAs nanowires with different crystal structures and different orientations are characterized for the first time, and the corresponding relation between the electrical transport properties and the crystal structures and orientations of nanowires is established.
3. **Effect of growth methods on the electrical transport properties of InAs nanowires**. Several growth systems of III–V materials have been developed, such as MBE system that grows materials by atomic layer epitaxy, MOCVD that equips with precision gas flow control systems, chemical beam epitaxy (CBE) and chemical vapor deposition (CVD) etc. Different growth methods have their own characteristics in doping, growth rate and cost. Therefore, studying the effects of different growth methods on the electrical transport properties of InAs nanowires is instructive for selecting future commercial growth systems. In this thesis, the differences in ON–OFF characteristics between MBE and MOCVD grown InAs nanowires are studied. And the changes on the performance of InAs

nanowire FETs that are brought by the indefinite concentration of background carbon doping in the growth procedure of nanowires in MOCVD system are also studied and supported with finite element simulation.

Research work in this thesis is focused on the above aspects, and the contents are arranged as follows: in Chap. 2, the experimental equipment, device fabrication process and extraction methods of basic electrical parameters used in this thesis will be introduced; in Chap. 3, we focus on the ultrahigh ON–OFF ratio that is caused by the larger bandgap due to quantum confinement effects in ultrathin InAs nanowires with diameter less than 10 nm; in Chap. 4, we presents the characterization of the electrical transport properties of InAs nanowires with different crystal structures and different crystal orientations, and combined with the variable temperature measurement and current thermal emission model, the effects of crystal structure and crystal orientation on the electrical transport properties of InAs nanowires are studied; in Chap. 5, we present the effect of carbon doping introduced during MOCVD growth of InAs nanowires on device performance. Then, all the aforementioned researches on InAs nanowire electronic device are summarized and forecasted at the end of thesis.

References

1. Ferain I, Colinge CA, Colinge JP (2011) Multigate transistors as the future of classical metal-oxide-semiconductor field-effect transistors. Nature 479(7373):310–316
2. Frank DJ (2002) Power-constrained CMOS scaling limits. IBM J Res Dev 46(2.3):235–244
3. Iwai H (2009) Roadmap for 22 nm and beyond (Invited Paper). Microelectron Eng 86(7–9):1520–1528
4. Theis TN, Solomon PM (2010) In quest of the "next switch" prospects for greatly reduced power dissipation in a successor to the silicon field-effect transistor. Proc IEEE 98(12):2005–2014
5. Riel H, Wernersson L-E, Hong M et al (2014) III–V compound semiconductor transistors—from planar to nanowire structures. MRS Bull 39(08):668–677
6. Kim DH, del Alamo JA, Antoniadis DA et al (2009) Extraction of virtual-source injection velocity in sub-100 nm HFETs. IEDM 09:861–864
7. Khakifirooz A, Antoniadis DA (2008) MOSFET performance scaling—Part I: historical trends. IEEE Trans Electron Devices 55(6):1391–1440
8. Kim D-H, Del Alamo JA (2010) 30-nm InAs PHEMTs with $f_T = 644$ GHz and $f_{max} = 681$ GHz. IEEE Electron Device Lett 31(8):806
9. Kim D-H, Del Alamo JA, Chen P et al (2010) 50-nm E-mode In0.7Ga0.3As PHEMTs on 100-mm InP substrate with fmax > 1 THz IEDM 10:692–695
10. del Alamo JA (2011) Nanometre-scale electronics with III–V compound semiconductors. Nature 479(7373):317–323
11. Waldron N, KimJ D-H, Del Alamo A (2010) A self-aligned InGaAs HEMT architecture for logic applications. IEEE Trans Electron Devices 57(1):297–304
12. Kim D-H, Del Alamo JA (2008) 30-nm InAs pseudomorphic HEMTs on an InP substrate with a current-gain cutoff frequency of 628 GHz. IEEE Electron Device Lett 29(8):830–833
13. Fukui T, Yoshimura M, Nakai E et al (2012) Position-controlled III-V compound semiconductor nanowire solar cells by selective-area metal-organic vapor phase epitaxy. Ambio 41:119–124

14. Tomioka K, Yoshimura M, Fukui T (2012) A III-V nanowire channel on silicon for high-performance vertical transistors. Nature 488(7410):189–192
15. Lee C-W, Yun S-R-N, Yu C-G et al (2007) Device design guidelines for nano-scale MuGFETs. Solid-State Electron 51(3):505–510
16. Auth CP, Plummer JD (1997) Scaling theory for cylindrical, fully-depleted, surrounding-gate MOSFET's. IEEE Electron Device Lett 18(2):74–76
17. Dick KA, Thelander C, Samuelson L et al (2010) Crystal phase engineering in single InAs nanowires. Nano Lett 10(9):3494–3499
18. Bao J, Bell DC, Capasso F et al (2009) Nanowire-induced Wurtzite InAs thin film on Zinc-Blende InAs substrate. Adv Mater 21(36):3654–3658
19. Crain J, Piltz RO, Ackland GJ et al (1995) Erratum: tetrahedral structures and phase transitions in III-V semiconductors. Phys Rev B 52(23):16936
20. Glas F, Harmand JC, Patriarche G (2007) Why does wurtzite form in nanowires of III-V zinc blende semiconductors? Phys Rev Lett 99(14):146101
21. Koguchi M, Kakibayashi H, Yazawa M et al (1992) Crystal structure change of GaAs and InAs whiskers from zinc-blende to wurtzite type. Jpn J Appl Phys 31(7):2061
22. Pan D, Fu M, Yu X et al (2014) Controlled synthesis of phase-pure InAs nanowires on Si(111) by diminishing the diameter to 10 nm. Nano Lett 14(3):1214–1220
23. Dubrovskii VG, Sibirev NV (2008) Growth thermodynamics of nanowires and its application to polytypism of zinc blende III-V nanowires. Phys Rev B 77(3)
24. Zhang Z, Lu Z-Y, Chen P-P et al (2013) Quality of epitaxial InAs nanowires controlled by catalyst size in molecular beam epitaxy. Appl Phys Lett 103(7):073109
25. Sladek K, Winden A, Wirths S et al (2012) Comparison of InAs nanowire conductivity: influence of growth method and structure. Phys Status Solidi (c) 9(2):230–234
26. Thelander C, Caroff P, Plissard S et al (2011) Effects of crystal phase mixing on the electrical properties of InAs nanowires. Nano Lett 11(6):2424–2429
27. Zhang Z, Zheng K, Lu ZY et al (2015) Catalyst Orientation-induced growth of defect-free zinc-blende structured <001> InAs Nanowires. Nano Lett 15(2):876–882
28. Xu H, Wang Y, Guo Y et al (2012) Defect-free <110> zinc-blende structured InAs nanowires catalyzed by palladium. Nano Lett 12(11):5744–5749
29. Hjort M, Lehmann S, Knutsson J et al (2014) Electronic and structural differences between Wurtzite and Zinc Blende InAs nanowire surfaces: experiment and theory. ACS Nano 8(12):12346–12355
30. Ning F, Tang L-M, Zhang Y et al (2013) First-principles study of quantum confinement and surface effects on the electronic properties of InAs nanowires. J Appl Phys 114(22):224304
31. Trägårdh J, Persson AI, Wagner JB et al (2007) Measurements of the band gap of wurtzite $InAs_{1-x}P_x$ nanowires using photocurrent spectroscopy. J Appl Phys 101(12):123701
32. De A, Pryor CE (2010) Predicted band structures of III-V semiconductors in the wurtzite phase. Phys Rev B 81(15):155210
33. Mead C, Spitzer W (1963) Fermi level position at semiconductor surfaces. Phys Rev Lett 10(11):471–472
34. Noguchi M, Hirakawa K, Ikoma T (1991) Intrinsic electron accumulation layers on reconstructed clean InAs(100) surfaces. Phys Rev Lett 66(17):2243–2246
35. Olsson LO, Andersson CBM, Hakansson MC et al (1996) Charge accumulation at InAs surfaces. Phys Rev Lett 76(19):3626–3629
36. Razavieh A, Mohseni PK, Jung K et al (2014) Effect of diameter variation on electrical characteristics of Schottky Barrier Indium Arsenide nanowire field-effect transistors. ACS Nano 8(6):6281–6287
37. Shi T, Fu M, Pan D et al (2015) Contact properties of field-effect transistors based on indium arsenide nanowires thinner than 16 nm. Nanotechnology 26(17):175202
38. Wagner RS, Ellis WC (1964) Vapor-liquid-solid mechanism of single crystal growth. Appl Phys Lett 4(5):89
39. Björk MT, Ohlsson BJ, Thelander C et al (2002) Nanowire resonant tunneling diodes. Appl Phys Lett 81(23):4458

40. Tomioka K, Izhizaka F, Fukui T (2015) Selective-area growth of InAs nanowires on Ge and vertical transistor application. Nano Lett 15(11):7253–7257
41. Dey AW, Thelander C, Lind E et al (2012) High-performance InAs nanowire MOSFETs. IEEE Electron Device Lett 33(6):791–793
42. Johansson S, Memisevic E, Wernersson L-E et al (2014) High-frequency gate-all-around vertical InAs nanowire MOSFETs on Si substrates. IEEE Electron Dev Lett 35(518–520)
43. Abay S, Persson D, Nilsson H et al (2014) Charge transport in InAs nanowire Josephson junctions. Phys Rev B 89(21)
44. Anindya Das YR, Most Y, Oreg Y et al. (2012) Zero-bias peaks and splitting in an Al–InAs nanowire topological superconductor as a signature of Majorana fermions. Nat Phys 8:887–895
45. Abay S, Nilsson H, Wu F et al (2012) High critical-current superconductor-InAs nanowire-superconductor junctions. Nano Lett 12(11):5622–5625
46. Fan D, Li S, Kang N et al (2015) Formation of long single quantum dots in high quality InSb nanowires grown by molecular beam epitaxy. Nanoscale 7(36):14822–14828
47. Miao J, Hu W, Guo N et al (2014) Single InAs nanowire room-temperature near-Infrared photodetector. ACS Nano 8(4):3628–3635
48. Shin HW, Lee SJ, Kim DG et al (2015) Short-wavelength infrared photodetector on Si employing strain-induced growth of very tall InAs nanowire arrays. Sci Rep 5:10764
49. Liu Z, Luo T, Liang B et al (2013) High-detectivity InAs nanowire photodetectors with spectral response from ultraviolet to near-infrared. Nano Res 6(11):775–783
50. Wei W, Bao X-Y, Soci C et al (2009) Direct heteroepitaxy of vertical InAs nanowires on Si substrates for broad band photovoltaics and photodetection. Nano Lett 9(8):2926–2934
51. Dey AW, Svensson J, Borg BM et al (2012) Single InAs/GaSb nanowire low-power CMOS inverter. Nano Lett 12(11):5593–5597
52. Svensson J, Dey AW, Jacobsson D et al (2015) III-V nanowire complementary metal-oxide semiconductor transistors monolithically integrated on Si. Nano Lett 15(12):7898–7904
53. Gu JJ, Neal AT, Ye PD (2011) Effects of $(NH_4)2_S$ passivation on the off-state performance of 3-dimensional InGaAs metal-oxide-semiconductor field-effect transistors. Appl Phys Lett 99(15):152113
54. Radosavljevic M, Chu-Kung B, Corcoran S et al (2009) Advanced high-K gate dielectric for high-performance short-channel $In_{0.7}Ga_{0.3}As$ quantum well field effect transistors on silicon substrate for low power logic applications. IEDM 9:319
55. Bryllert T, Wernersson L-E, Löwgren T et al (2006) Vertical wrap-gated nanowire transistors. Nanotechnology 17(11):S227–S230
56. Rehnstedt C, Martensson T, Thelander C et al (2008) Vertical InAs nanowire wrap gate transistors on Si substrates. IEEE Trans Electron Devices 55(11):3037–3041
57. Thelander C, FrobergFroberg LE, Rehnstedt C et al (2008) Vertical enhancement-mode InAs nanowire field-effect transistor with 50-nm wrap gate. IEEE Electron Device Lett 29(3):206–208
58. Egard M, Johansson S, Johansson AC et al (2010) Vertical InAs nanowire wrap gate transistors with $f_t > 7$ GHz and $f_{max} > 20$ GHz. Nano Lett 10(3):809–812
59. Berg M, Persson KM, Wu J et al (2014) InAs nanowire MOSFETs in three-transistor configurations: single balanced RF down-conversion mixers. Nanotechnology 25(48):485203
60. Persson K-M, Berg M, Borg MB et al (2013) Extrinsic and intrinsic performance of vertical InAs nanowire MOSFETs on Si substrates. IEEE Trans Electron Devices 60(9):2761–2767
61. Tomioka K, Fukui T (2011) Tunnel field-effect transistor using InAs nanowire/Si heterojunction. Appl Phys Lett 98(8):083114
62. Tanaka T, Tomioka K, Hara S et al (2010) Vertical surrounding gate transistors using single InAs nanowires grown on Si substrates. Appl Phys Express 3(2):5003
63. Nam S, Jiang X, Xiong Q et al (2009) Vertically integrated, three-dimensional nanowire complementary metal-oxide-semiconductor circuits. Proc Natl Acad Sci 106(50):21035–21038
64. Dick KA, Deppert K, Samuelson L et al (2006) Optimization of Au-assisted InAs nanowires grown by MOVPE. J Cryst Growth 297(2):326–333

65. Ghalamestani SG, Johansson S, Borg BM et al (2012) Uniform and position-controlled InAs nanowires on 2″ Si substrates for transistor applications. Nanotechnology 23(1):015302
66. Tomioka K, Mohan P, Noborisaka J et al (2007) Growth of highly uniform InAs nanowire arrays by selective-area MOVPE. J Cryst Growth 298:644–647
67. Dimakis E, Lähnemann J, Jahn U et al (2011) Self-assisted nucleation and vapor-solid growth of InAs nanowires on bare Si(111). Cryst Growth Des 11(9):4001–4008
68. Persson KM, Sjöland H, Lind E et al (2014) InAs nanowire MOSFET differential active mixer on Si-substrate. Electron Lett 50(9):682–683
69. Ford AC, Chuang S, Ho JC et al (2010) Patterned p-doping of InAs nanowires by gas-phase surface diffusion of Zn. Nano Lett 10(2):509–513
70. Ionescu AM, Riel H (2011) Tunnel field-effect transistors as energy-efficient electronic switches. Nature 479(7373):329–337
71. De Michielis L, Lattanzio L, Moselund K et al (2013) Tunneling and occupancy probabilities: how do they affect tunnel-FET behavior? IEEE Electron Device Lett 34(6):726–728
72. Borg M, Schmid H, Moselund KE et al (2014) Vertical III-V nanowire device integration on Si(100). Nano Lett 14(4):1914–1920
73. Dey AW, Svensson J, Ek M et al (2013) Combining axial and radial nanowire heterostructures: radial esaki diodes and tunnel field-effect transistors. Nano Lett 13(12):5919–5924
74. Dey AW, Borg BM, Ganjipour B et al (2013) High-current GaSb/InAs (Sb) nanowire tunnel field-effect transistors. IEEE Electron Device Lett 34(2):211–213
75. Dayeh SA, Aplin DP, Zhou X et al (2007) High electron mobility InAs nanowire field-effect transistors. Small 3(2):326–332
76. Zhao Y, Candebat D, Delker C et al (2012) Understanding the impact of Schottky barriers on the performance of narrow bandgap nanowire field effect transistors. Nano Lett 12(10):5331–5336
77. Appenzeller J, Radosavljević M, Knoch J et al (2004) Tunneling versus thermionic emission in one-dimensional semiconductors. Phys Rev Lett 92(4)
78. Oktyabrsky S, Ye PD (2010) Fundamentals of III-V semiconductor MOSFETs. Springer, New York
79. Sze SM, Ng KK (2007) Physics of semiconductor devices. Wiley, Hoboken
80. Takei K, Fang H, Kumar SB et al (2011) Quantum confinement effects in nanoscale-thickness InAs membranes. Nano Lett 11(11):5008–5012
81. George SM (2010) Atomic layer deposition: an overview. Chemical Review 110:111–131
82. Saint-Cast P, Kania D, Hofmann M et al (2009) Very low surface recombination velocity on p-type c-Si by high-rate plasma-deposited aluminum oxide. Appl Phys Lett 95(15):151502
83. Wang Z, Xu H, Zhang Z et al (2010) Growth and performance of yttrium oxide as an ideal high-kappa gate dielectric for carbon-based electronics. Nano Lett 10(6):2024–2030
84. Wallace RM, McIntyre PC, Kim J et al (2009) Atomic layer deposition of dielectrics on Ge and III–V materials for ultrahigh performance transistors. MRS Bull 34(07):493–503
85. Takei K, Kapadia R, Fang H et al (2013) High quality interfaces of InAs-on-insulator field-effect transistors with ZrO_2 gate dielectrics. Appl Phys Lett 102(15):153513
86. Wheeler D, Seabaugh A, Froberg L et al. (2007) Electrical properties of HfO_2/InAs MOS capacitors. In: Semiconductor device research symposium, pp 1–2
87. Mahata C, Byun YC, An CH et al. (2013) Comparative study of atomic-layer-deposited stacked (HfO_2/Al_2O_3) and nanolaminated ($HfAlO_x$) dielectrics on $In_{0.53}Ga_{0.47}As$. ACS Appl Mater & Interfaces 5(10):4195–4201
88. Ko H, Takei K, Kapadia R et al (2010) Ultrathin compound semiconductor on insulator layers for high-performance nanoscale transistors. Nature 468(7321):286–289
89. Tomioka K, Yoshimura M, Fukui T (2013) Sub 60 mV/decade switch using an InAs nanowire-Si heterojunction and turn-on voltage shift with a pulsed doping technique. Nano Lett 13(12):5822–5826
90. Hilner E, Hakanson U, Froberg LE et al (2008) Direct atomic scale imaging of III-V nanowire surfaces. Nano Lett 8(11):3978–3982

91. Ford AC, Ho JC, Chueh Y-L et al (2008) Diameter-dependent electron mobility of InAs nanowires. Nano Lett 9(1):360–365
92. Takahashi T, Takei K, Adabi E et al (2010) Parallel array InAs nanowire transistors for mechanically bendable, ultrahigh frequency electronics. ACS Nano 4(10):5855–5860
93. Esseni D, Pala MG (2013) Interface traps in InAs nanowire tunnel FETs and MOSFETs—Part II: comparative analysis and trap-induced variability. IEEE Trans Electron Devices 60(9):2802–2807
94. Hertenberger S, Rudolph D, Bichler M et al (2010) Growth kinetics in position-controlled and catalyst-free InAs nanowire arrays on Si(111) grown by selective area molecular beam epitaxy. J Appl Phys 108(11):114316
95. Dey AW (2013) Low-power nanowire circuits and transistors. Doctoral thesis, Lund University
96. Léonard F, Talin A (2006) Size-dependent effects on electrical contacts to nanotubes and nanowires. Phys Rev Lett 97(2):026804
97. Lind E, Persson MP, Niquet Y-M et al (2009) Band structure effects on the scaling properties of [111] InAs nanowire MOSFETs. IEEE Trans Electron Devices 56(2):201–205
98. Chuang S, Gao Q, Kapadia R et al (2013) Ballistic InAs nanowire transistors. Nano Lett 13(2):555–558
99. Dayeh SA, Susac D, Kavanagh KL et al (2009) Structural and room-temperature transport properties of Zinc Blende and Wurtzite InAs nanowires. Adv Func Mater 19(13):2102–2108
100. Ullah AR, Joyce HJ, Burke AM et al (2013) Electronic comparison of InAs wurtzite and zincblende phases using nanowire transistors. Phys Status Solidi—Rapid Res Lett 7(10):911–914
101. Shimoida K, Yamada Y, Tsuchiya H et al (2013) Orientational dependence in device performances of InAs and Si nanowire MOSFETs under ballistic transport. IEEE Trans Electron Devices 60(1):117–122
102. Alam K, Sajjad RN (2010) Electronic properties and orientation-dependent performance of InAs nanowire transistors. IEEE Trans Electron Devices 57(11):2880–2885
103. Boxberg F, Sondergaard N, Xu HQ (2012) Elastic and piezoelectric properties of zincblende and wurtzite crystalline nanowire heterostructures. Adv Mater 24(34):4692–4706
104. Li X, Wei X, Xu T et al (2015) Remarkable and crystal-structure-dependent piezoelectric and piezoresistive effects of InAs nanowires. Adv Mater 27(18):2852–2858
105. Zheng K, Zhang Z, Hu Y et al (2016) Orientation dependence of electromechanical characteristics of defect-free InAs nanowires. Nano Lett 16(3):1787–1793
106. Zhang X, Zou J, Paladugu M et al (2009) Evolution of epitaxial InAs nanowires on GaAs (111) B. Small 5(3):366–369
107. Joyce HJ, Gao Q, Tan HH et al (2007) Twin-free uniform epitaxial GaAs nanowires grown by a two-temperature process. Nano Lett 7(4):921–926
108. Woo RL, Xiao R, Kobayashi Y et al (2008) Effect of twinning on the photoluminescence and photoelectrochemical properties of indium phosphide nanowires grown on silicon (111). Nano Lett 8(12):4664–4669
109. Schroer MD, Petta JR (2010) Correlating the nanostructure and electronic properties of InAs nanowires. Nano Lett 10(5):1618–1622
110. Jiang X, Xiong Q, Nam S et al (2007) InAs/InP radial nanowire heterostructures as high electron mobility devices. Nano Lett 7(10):3214–3218
111. Caroff P, Wagner JB, Dick KA et al (2008) High-quality InAs/InSb nanowire heterostructures grown by metal-organic vapor-phase epitaxy. Small 4(7):878–882
112. Ullah AR, Joyce HJ, Burke AM et al (2013) Electronic comparison of InAs wurtzite and zincblende phases using nanowire transistors. Phys Status Solidi – Rapid Res Lett **7**(10):911–914
113. Takei K, Chuang S, Fang H et al (2011) Benchmarking the performance of ultrathin body InAs-on-insulator transistors as a function of body thickness. Appl Phys Lett 99(10):103507

Chapter 2
Fabrication, Characterization and Parameter Extraction of InAs Nanowire-Based Device

Abstract Nanofabrication, low noise electrical measurement and various nanoscale characterization methods are frequently used in the study. This chapter will give as introduction on the growth method, fabrication techniques, characterization methods of materials and devices, measurement systems, and way to extract the electrical parameters of InAs nanowires FET devices.

Keywords Fabrication of nanowire devices · Electrical measurement
Parameter extraction · Structural characterization

Nanofabrication, low noise electrical measurement and various nanoscale characterization methods are frequently used in the study of InAs nanowire-based devices. This chapter will give an introduction on the growth method, fabrication techniques, characterization methods of materials and devices, measurement systems, and way to extract the electrical parameters of InAs nanowires FET devices.

2.1 Growth of InAs Nanowires

There are two ways to obtain InAs nanowires, which are typically called "top-down" and "bottom-up", respectively. All nanowires used in this work are fabricated by the "bottom-up" method. For each method, the grown nanowires have different pros and cons. Among various growth methods, MBE and MOCVD become the most favorable systems for the III-V compound semiconductors due to their advantage of high controllability of crystal and heterostructures [1–4]. In this work, we use nanowires grown by both MBE and MOCVD. And both these two kinds of nanowires are grown on the Si (111) substrate without intentionally doping during their growth.

MBE is a semiconductor growth method derived from the vacuum evaporation technology of the 1950s. By heating an ultrahigh-purity source element, a molecule flow is evaporated onto a semiconductor substrate in the chamber after reaching a certain temperature. A schematic diagram of the growth of III-V compound nanowires in MBE system is shown in Fig. 2.1 [5]. By MBE growth method the thickness of

M. Fu, *Electrical Properties of Indium Arsenide Nanowires and Their Field-Effect Transistors*, Springer Theses, https://doi.org/10.1007/978-981-13-3444-3_2

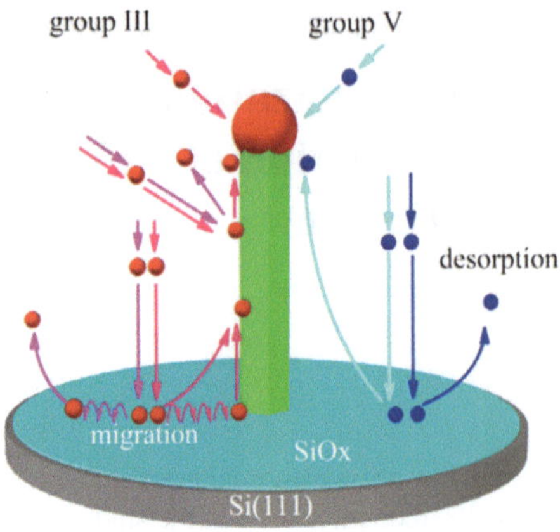

Fig. 2.1 Schematic diagram of growing III-V compound nanowires in MBE system. In the figure, group III flux and group V flux are the molecular flows generated by heating high-purity sources. During the growth process, two elemental molecular flows are alternately released into the high vacuum chamber, and the III-V nanowires are grown layer by layer. Adapted with permission from [5]. Copyright 2013 American Chemical Society

the sample can be precisely controlled to monolayer, and because the chamber for growth is in an ultra-high vacuum condition, it can strictly control the component of the epitaxial layer to avoid unintentional doping. Therefore, the MBE system is very suitable to the research on intrinsic properties and novel heterostructure of semiconductors. But the MBE method also has drawbacks such as very slow growth rate. Thus, most of the commercially available III-V compound semiconductor are grown by MOCVD.

MOCVD uses organic compound of group III element and hydride of group V element as growth sources, and performs vapor-phase epitaxy on a substrate by a thermal decomposition reaction. Figure 2.2 shows a schematic diagram of growing III-V compound nanowires by an MOCVD system [6]. Compared with the MBE system, MOCVD system has higher efficiency, and through the sophisticated gas flow control system, MOCVD system is able to precisely control the rate of synthesis reaction and thus obtain nanowires with a particular composition, crystal structure and scale. However, since organic compounds are required in decomposition reaction during growth processes, it is unavoidable to introduce carbon as dopant [7].

The schematic diagram for the nanowire growth process by these two mechanisms are shown in Fig. 2.3 [8] and Fig. 2.4 [9], respectively. As for InAs nanowires, these two growth mechanisms each have their own advantages. By using metal particles as the catalyst to assist in the growth of nanowires, the large-scale synthesis of InAs nanowires with high-quality crystal structures is possible at this stage [10, 11]. However, since most of the metals (including Au) used for catalysis are easy to diffuse into the Si substrate or III-V nanowires, these impurities of metal atoms might greatly affect the properties of III-V material (especially optical properties).

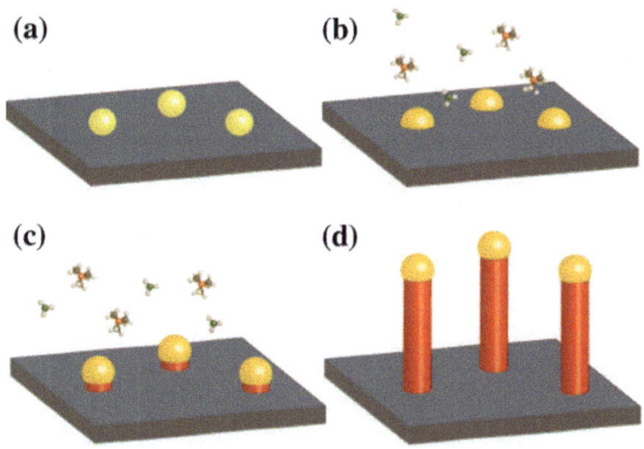

Fig. 2.2 A schematic diagram of growing III-V nanowires in MOCVD system. **a** depositing metal particles on growth substrate as catalyst; **b** heating the growth substrate, while introducing the compound gas of Group III and V into the growth chamber to form alloy particles with the metal particles; **c** the III and V components in the liquid alloy particles have reached the saturation state, and the nucleation occurs at the interface between alloy particle and growth substrate, and then nanowires begin to grow. **d** continuously growing the III-V nanowires [6]

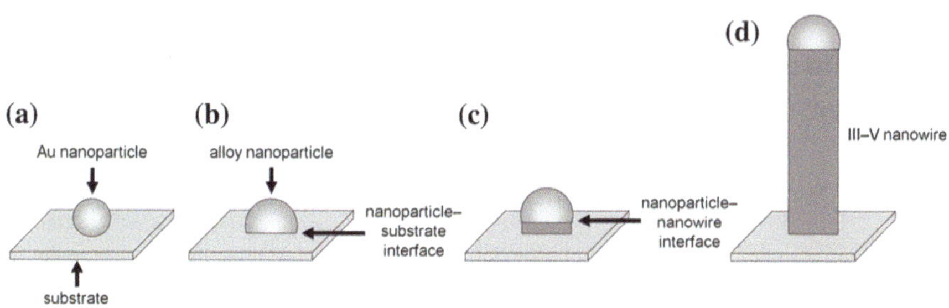

Fig. 2.3 Schematic diagram of reactions used to grow InAs nanowires with Au particle catalyst. **a** depositing Au particles on growth substrate; **b** introducing In and As gas to form co-solved alloy particles with metal particles; **c** beginning to form InAs nanowires on the interface between co-solved alloy particles and growth substrate; **d** the continuous growth of nanowires on the surface between co-solved alloy particles and nanowires through chemical reaction. Reprinted from Ref. [8], Copyright 2011 with permission from Elsevier

In this regard, growing nanowires with self-catalyzed mechanisms may help to solve this problem because of an inherent catalyst-free mechanism.

As a result, nanowires used in different chapters will differ in the aspects of growth systems and growth mechanisms, according to different research purposes. Specifically, the nanowires used in Chaps. 3 and 4 are all grown by MBE. In Chap. 5, the InAs nanowires are synthesized using self-catalyzed growth mechanism in both MBE and MOCVD systems. Details of the growth conditions and crystal structure of InAs nanowires will be respectively described in each chapter.

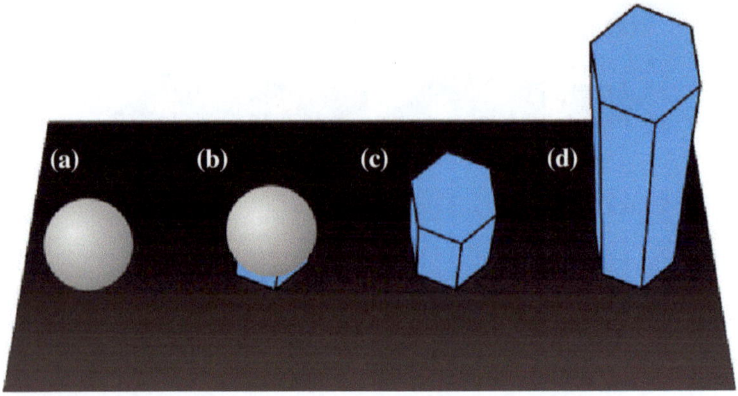

Fig. 2.4 Schematic diagram of self-catalyzed growth of InAs nanowires. **a** introducing the reactant of Into form In droplets on the surface of growth substrate. **b** introducing the reactant of As and then beginning to grow InAs nanowires by using In droplets as nucleation centers. **c** as reacts with In droplets and In droplet at the top of nanowires gradually disappear. **d** InAs nanowires grow in vapor solid mode under As-rich conditions. Reprinted with the permission from Ref. [9]. Copyright 2011 American Chemical Society

2.2 Characterization, Device Fabricating and Electrical Measurement Equipment of Nanowires

2.2.1 Characterization Equipment

Optical microscopy

In this work, optical microscopy is used to quickly examine the nanowires' geometry and coarsely locate the position. It is also used in multiple micro/nano fabrication processes involved in this work. The apparatus used in this work is Carl Zeiss Axio Imager (as shown in Fig. 2.5a).

Scanning electron microscopy

The dimensions of nanomaterials and devices in this work are frequently smaller than the diffraction limit of visible light, and exceed the resolution of optical microscopy. Therefore, in order to further characterize the specific shape and scale of nanomaterials and devices, it is necessary to use the equipment having higher resolution (at nanometer scale). In this work, we use scanning electron microscopy (SEM) to conduct the aforementioned characterization. The SEM uses electron beams with energy ranging from 1 to 30 keV for imaging, and their resolution limit can reach several nanometers. Their image contrast is formed by detecting the signals of secondary electrons and backscattered electrons activated by high-energy electron beam on the surface of the sample. The SEM used in this work is field-emission SEM, and its product model is FEI Quanta 600F (as shown in Fig. 2.5b). In this work, SEM is

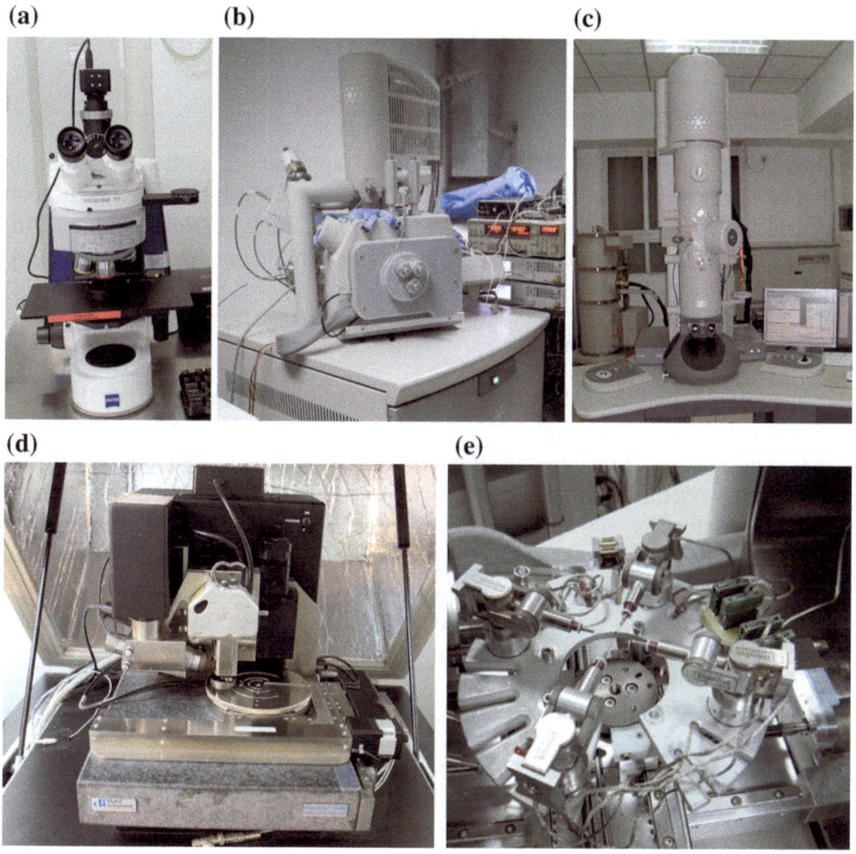

Fig. 2.5 **a** photograph of the optical microscopy apparatus of Carl Zeiss Axio Imager; **b** photograph of the environmental SEM of FEI Quanta 600F; **c** photograph of the field-emission high resolution TEM of FEI Tecnai F20; **d** photograph of the AFM of Veeco Multimode V; **e** photograph of the nano-manipulation system apparatus of Kleindiek MM3A

mainly used to choose growth substrate of InAs nanowires, observe geometry of InAs nanowires, accurately locate InAs nanowires, observe the topography of devices.

Transmission electron microscopy

Transmission electron microscopy (TEM) provides spatial resolution at atomic scale. It mainly relies on the interaction between coherent electron beams and the sample to generate transmitted electrons with information of the sample for imaging. The TEM used in this work is a FEI Tecnai F20 field-emission TEM with an acceleration voltage of 200 kV (as shown in Fig. 2.5c) with a line limit resolution of 1 Å and a dot resolution of 2.4 Å. In this work, the device is mainly used to determine the crystal structure and morphology of InAs nanowires. Moreover, the characteristic X-rays emitted by nanowires under electron beam irradiation are analyzed through a attached energy dispersive X-ray spectroscopy (EDS), in order to determine the relative content of elements in InAs nanowires.

Atomic force microscopy

Atomic force microscopy (AFM) is one of the apparatuses used to measure the height of material surface. In the direction of height, its resolution can reach 1 nm. The model of AFM used in this work is Veeco Multimode V (as shown in Fig. 2.5d). And this apparatus is mainly used to accurately determine the diameter of the InAs nanowires.

Nano-manipulation system

The nano-manipulation system used in this work is home-built in the SEM, and the system model is Kleindiek MM3A (shown in Fig. 2.5f), which is a tool for in situ manipulating the nanowires at nanometer scale. The system consists of four independently operational nanoprobes, each of which contains a robotic arm that can move in three-dimensional spherical coordinates. Driven by piezoelectric ceramics, the arm can move in one radial and two angular directions with a maximum movement accuracy of 0.5 nm. In this work, nano-manipulation techniques are mainly used to transfer InAs nanowires that have completed electrical measurements to a coordinate microgrid used for TEM characterization.

2.2.2 Apparatus for Device Fabrication

Ultraviolet exposure system

Ultraviolet (UV) exposure is a traditional method used to transfer device patterns in semiconductor processes. A wafer coated with a layer of photosensitive polymer material (UV photoresist) is placed under a mask that covers the ultraviolet light in specified areas, and then by controlling the areas of photochemical decomposition or polymerization reactions, the patterns on masks can be transferred to the photoresist. The product model of the UV exposure system used in this work is Süss MJB4 (as shown in Fig. 2.6a). This apparatus can perform exposure on wafers with diameter up to 4 inches, and produce a minimal line width of 1 μm. In this work, the UV exposure system is mainly used to fabricate large marks for locating the position of nanowires and their devices.

Electron-beam lithography

Electron-beam lithography (EBL) is also used to pattern the resist layer upon the wafer. Its main difference from the UV exposure system is that on one hand, its light source is electron beam having energy of several thousand electron volts (keV) to several tens of keV; on the other hand, the photosensitive material used (collectively

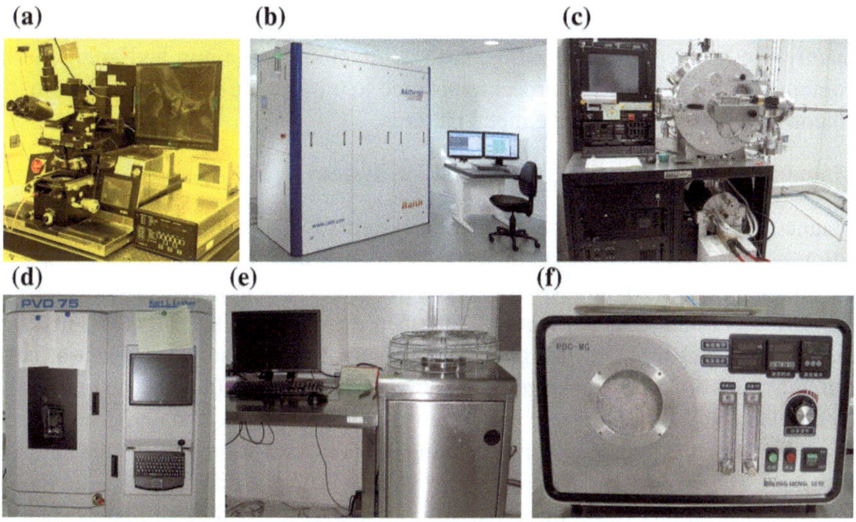

Fig. 2.6 **a** photo of the UV exposure system of Süss MJB4; **b** photo of the electron beam exposure system of Raith 150 II; **c** photo of the electron-beam evaporation of Kurt J. Lesker AXXIS; **d** photo of the magnetron sputtering thin film deposition system of Kurt J. Lesker PVD75; **e** photo of the atomic layer deposition system of Cambridge Nanotech Savannah 100; **f** photo of oxygen plasma cleaning equipment of Mingheng PDC-MG

referred to as electron beam resist) is the polymer sensitive to electron beams. Because the wavelength of high-energy electron beam is much smaller than that of the UV light, the minimal line width that can be achieved in electron-beam system is much smaller than that of the UV exposure system. The electron-beam lithography used in this work is a direct-write Raith 150 Two electron-beam lithography (as shown in Fig. 2.6b). It can achieve a minimal linewidth of about 10 nm and maximally process the 4' wafer. The function of direct-write exposure eliminates the need of masks, and increases the flexibility and variability of graphics transfer, so that makes it suitable for fabricating complex and diverse nanoscale devices. In this work, the electrodes of the device were patterned by this system.

Electron-beam evaporation

Electron-beam Evaporation (EBE) in this work is mainly used to deposit metal films. This apparatus uses the following steps to deposit the metal films. First, the targeted material in the crucible is heated by a high-energy electron beam to form a high temperature region. Then, the material in this region is melted and evaporated onto the surface of the targeted substrate. The film formed by electron-beam deposition has the characteristics of good collimation and small damage to the target substrate. The product model of apparatus used in this work is the Kurt J. Lesker AXXIS thin-film deposition apparatus as shown in Fig. 2.6c. In this work, this apparatus is mainly used to deposit metal films, such as titanium, chromium and gold, and in combination with the lift-off process, it is used to transfer the designed pattern to the silicon substrate and obtain the metal patterns.

Magnetron sputtering thin film deposition system

Magnetron sputtering thin film deposition system (PVD) is another thin-film deposition system used in this work. Its working mechanism is to bombard the surface of the targeted material by high-energy argon plasma, so that atoms are separated from the surface of bulk targeted material and then deposited on the targeted substrate. Due to the bombardment of high-energy plasma, atoms of targeted material have different emission directions and relatively high energy, and thus the step coverage of the deposited film is better; and because atoms of targeted material have relatively high energy, they have some cleaning effect as well as damage on the targeted substrate. The product model used in this work is Kurt J. Lesker PVD75 (as shown in Fig. 2.6d). In this work, the apparatus is mainly used to deposit metals such as chromium and gold and to clean native oxides of III-V nanowires at the same time.

Atomic layer deposition

Atomic layer deposition (ALD) is one thin-film deposition system with excellent step coverage and precise control on film thickness. The working principle is as follows: the reaction gas (precursor) is separately introduced into the reaction chamber in multiple times, a single layer of molecular is formed on the surface of the targeted substrate each time by diffusing a small quantity of precursor, and thus the atomic-level reaction is realized. Through this process, the obtained films have a particular thickness and few defects. The product model used in this work is the Cambridge Nanotech Savannah 100 (as shown in Fig. 2.6e). The deposited aluminum oxide or yttrium oxide film is mainly used as the gate dielectric of devices and can also be used for the package of devices and surface treatment of III-V material.

Oxygen plasma cleaning equipment

Oxygen plasma cleaning is an organic cleaning procedure commonly used in semiconductor processes. The oxygen plasma generated by high voltage physically bombards and chemically etches the organics at the same time, therefore removing organic molecules including the UV photoresist and electron-beam resist. The product model used in this work is Mingheng PDC-MG (as shown in Fig. 2.6f). It is mainly used to remove the photoresist which functions as the support layer when fabricates suspended InAs nanowire devices and to etch the residual resist in EBL process.

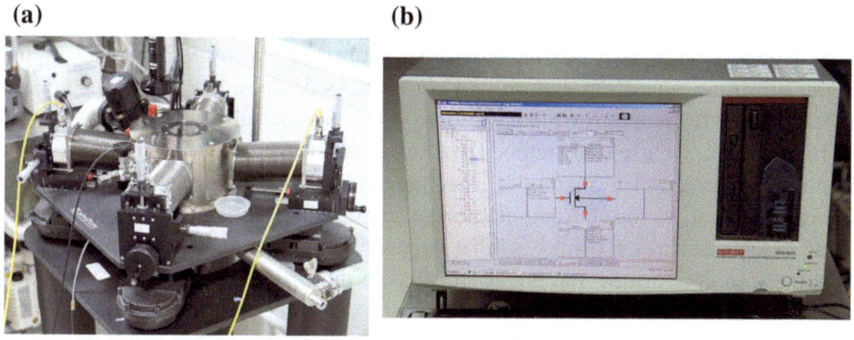

Fig. 2.7 **a** photo of the cryogenic probe station of Lakeshore TTP4; **b** photo of semiconductor parameter analyzer of Keithly 4200SCS

2.2.3 Apparatus to Characterize the Electrical Properties of Devices

Cryogenic probe station

Using a probe station to measure the electrical properties of micro/nano-scale electronic devices has the advantages of high flexibility and efficiency. The probe station used in this work is LakeShore TTP4 shown in Fig. 2.7a. This apparatus is equipped four movable probes whose tip can have minimal diameter of 20 μm. Through the consumable cooling module and temperature control system of the station, it can realize variable-temperature DC electrical test ranging from 4.2 K to room temperature.

Semiconductor parameter analyzer

The electrical measurement system used in this work is the semiconductor parameter analyzer of the Keithley 4200SCS as shown in Fig. 2.7a. The measurement system's accuracy of minimal current measurement is up to fA, and there are four measurement ports that can output and simultaneously measure current and voltage. Therefore, it can satisfy the need for electrical measurement of a single FET.

2.3 Fabrication Processes of InAs Nanowires Devices

2.3.1 Dispersion and Transfer of InAs Nanowires

In order to evaluate the relevant electrical parameters of InAs nanowires with different crystallographic parameters, and growth conditions, InAs nanowires are fabricated as planar devices for electrical measurements. Since the InAs nanowires in this work are all epitaxially grown on the (111) Si wafer by MBE or MOCVD, the epitaxial growth direction is mostly perpendicular to the substrate or at a specific angle relative to the substrate. Therefore, before fabricating InAs devices, the nanowires need to be transferred from the growth substrate to other specific substrates first.

In this work, we use two methods to transfer nanowires: mechanical transfer and solution dispersion. The method of mechanically transferring nanowires is shown in Fig. 2.8. First, we use a triangular, cleanroom tissue tip to gently or evenly draw on an InAs nanowires substrate, so that the nanowires are detached from the growth substrate by electrostatic force and attached to the tip of the tissue; then, contact the tip of the ultra-clean paper with the device substrate on which the metal mark has been fabricated. Because the Si substrates are more electrostatically attractive to InAs nanowires, the nanowires will detach from the paper tip or break and lie flat on the device substrate. The method of mechanically transferring nanowires belongs to dry transfer process, and does not expose nanowires to solution or other substances that may contaminate the sample, so the transferred nanowires are relatively clean. Therefore, in this work, nanowires in most devices are transferred by using this method. However, because InAs nanowires on some growth substrates are too thin or too sparse, or because the surface of growth substrate has a large fluctuation, the InAs nanowires cannot be transferred onto the device substrate by mechanical transfer method. Under these circumstances, we use the solution to transfer nanowires. As shown in Fig. 2.9, first, we use acetone or alcohol to clean the backside of growth substrate (removing the possible contaminants on the backside of substrate), immerse the whole substrate into a small amount of alcohol solution in the centrifuge tube or other small-capacity container, and then make the nanowires detached from the growth substrate by low-power ultrasonication; after a few seconds, take several drops from the upper part of the solution, drip them onto the device substrate with metal mark, and dry it naturally. Due to the large amount of impurities in the droplets and possible organic or ionic contaminants, the device substrate with nanowires needs to be carefully cleaned with acetone, alcohol and deionized water before locating the nanowires. It is likely to introduce contamination to nanowires and substrates with this method, and the transferred nanowires are usually sparsely dispersed on the device substrate. Thus, if we use this method to transfer the nanowires, substrate cleaning and volume of solution used in ultrasonic process are two key parameters and needed to be precisely controlled. In this work, only small-diameter WZ-phase InAs nanowires grown along <0001> direction used in the Chap. 3 are transferred by solution dispersion.

2.3.2 General Fabrication Process for Planar InAs Nanowire Devices

In this work, the device substrate is typically a heavily doped Si (100) substrate with silicon dioxide (SiO_2) on its surface. Before transferring InAs nanowires to the

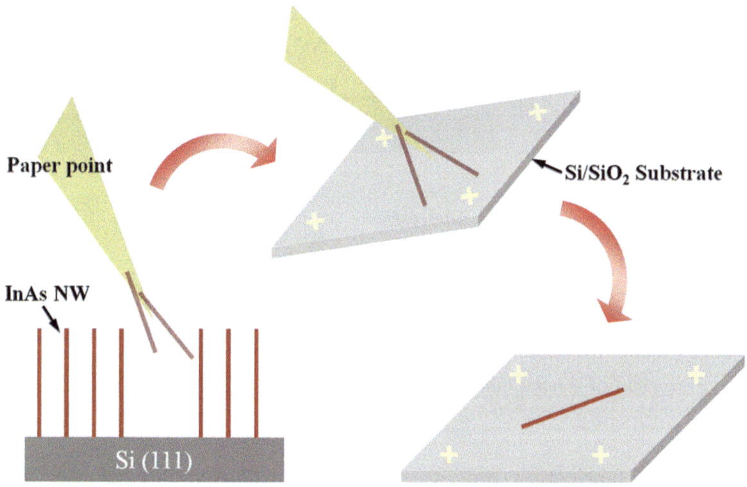

Fig. 2.8 Mechanically transfer the InAs nanowires from the growth substrate to the device substrate with the metal mark

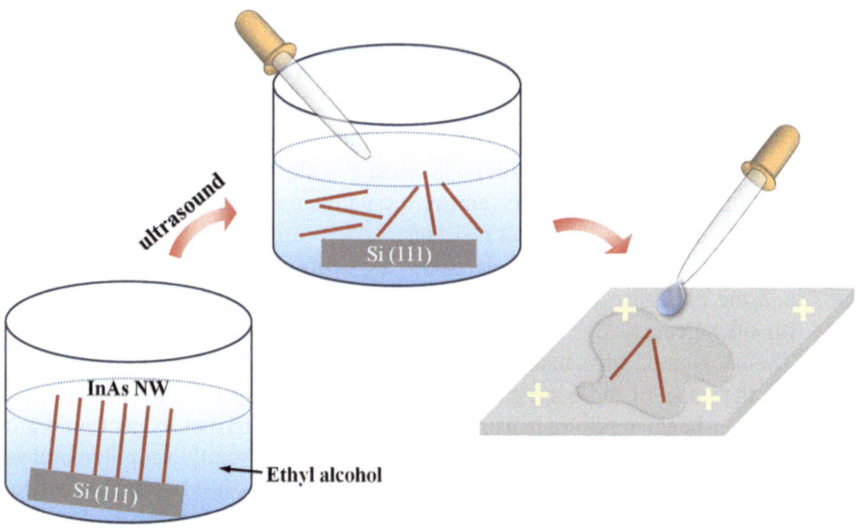

Fig. 2.9 A process of transferring InAs nanowires from a growth substrate to a device substrate with metal marks by solution

device substrate, Metal (Ti/Au) marks are fabricated by photolithography, electron beam exposure, and electron beam evaporation, in order to coarsely and precisely locating nanowires in the following steps. After transferring InAs nanowires onto the device substrate, back-gated or top-gated InAs nanowire planar FETs can be obtained through a series of micro/nano fabrication processes.

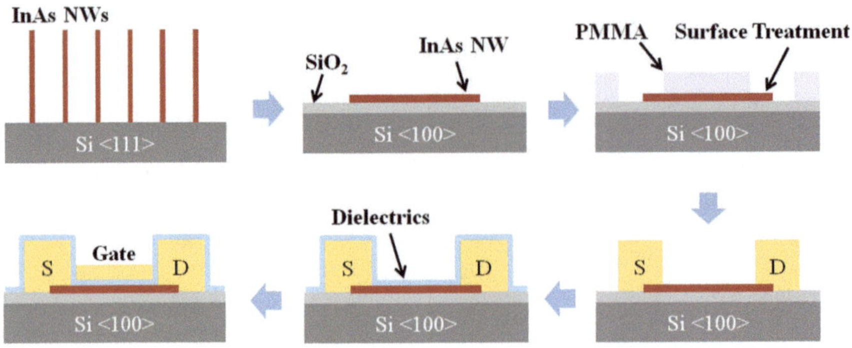

Fig. 2.10 Flow chart of general fabrication process of InAs nanowire planar device

The specific fabrication processes of nanowire-based devices are shown in Fig. 2.10:

(1) InAs nanowires are transferred (mechanically or by solution) and dispersed from growth substrate to Si/SiO_2 substrate with a metal marks. Here SiO_2 is thermal oxide layer of 300 nm.
(2) A layer of electron-beam exposure resist PMMA 950 K (MicroChem) is spin-coated onto the Si/SiO_2 substrate, and the designed source-drain electrode pattern is formed on the electron-beam exposure.
(3) The surface of InAs nanowires is chemically or physically cleaned by $(NH_4)_2S_x$ solution or argon plasma to remove the native oxide layer on the surface, in order to decrease the contact resistance.
(4) The source and drain metal electrodes are deposited by EBE or PVD system, and the excessive metal and electron-beam exposure resist are removed by lift-off process.
(5) The back-gated InAs nanowire devices are completed.
(6) On the basis of the back-gated devices, a double-layer resists composed of electron-beam exposure resists having different sensitivities to electron beams are spin-coated onto the surface of device substrate. Usually, a copolymer layer which is more sensitive to electron beam irradiation is used as the lower layer of bilayer resists, and a PMMA 950 K which is less sensitive is the upper layer. (As shown in Fig. 2.11). Then, the designed gate electrode pattern is transferred to the electron beam exposure resists by EBL system. (As shown in Fig. 2.11).
(7) The high-k gate oxide dielectric (Al_2O_3 or HfO_2) is deposited using ALD.
(8) Placing the sample directly into the EBE chamber for deposition of the gate electrode metal. Due to the high collimation of the EBE, the size of the metal-gate electrodes will be determined by the width of the smallest window in the electron beam exposure resists; while because the film deposited by ALD has excellent step coverage, the width of the gate dielectric is substantially equal to the lower surface window of the electron beam exposure resists. Consequently,

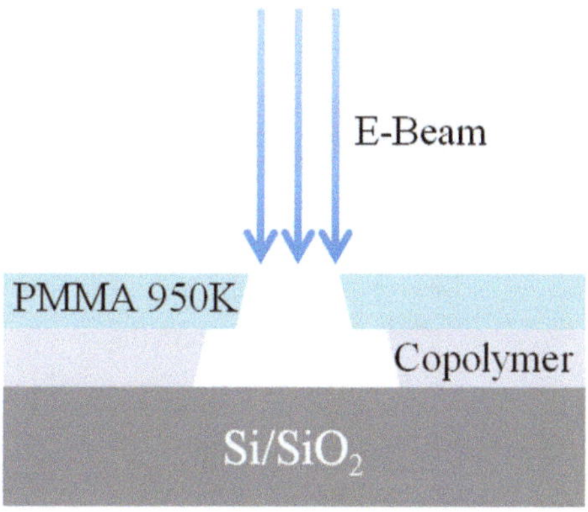

Fig. 2.11 Schematic diagram of the top-gate double-layer resists after E-beam irradiation

the width of the gate dielectric is generally wider than that of the electrode metal by about 150 nm (as shown in Fig. 2.11), which can effectively prevent the gate metal from contacting with the source and drain metal.

(9) Using the lift-off process to remove excessive metal, gate dielectric, and electron beam exposure resist.
(10) The top-gated InAs nanowire devices is completed.

The top-gated InAs nanowire devices fabricated by the above processes will be used in Chap. 3 for studying electrical parameters of the ultrathin InAs nanowires devices, and the back-gated devices will be used in Chap. 3 for studying the field-effect electron mobility and contact resistances of devices, and will also be used in Chap. 5 for studying the performance of devices.

2.3.3 Fabrication Process for Suspended InAs Nanowire Devices

Nanomaterials and their devices are often susceptible to the environment due to their large surface area to volume ratio. In order to eliminate the effects of substrate to contact and electrical transport of devices, we fabricate the suspended devices and do the electrical measurements. In addition, combined with nano-manipulation technology, we are able to transfer the InAs nanowires in suspended devices to micro grids for TEM characterization. Through this method, we sequentially perform the electrical characterization and the atomic-level structural characterization on the same InAs nanowires.

The specific fabrication processes of suspended nanowire-based devices are shown in Fig. 2.12:

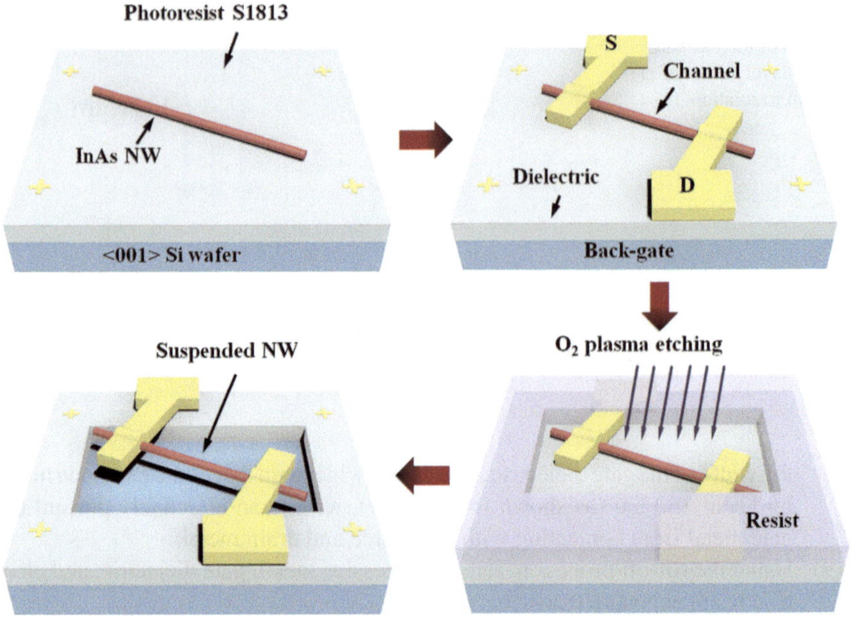

Fig. 2.12 Flow chart of the fabrication processes of suspended InAs nanowire device

(1) After spin coating a 1.2 μm UV photoresist S1813 on a heavily doped Si (100) substrate, cure the Si substrate on a hot plate at 200 °C for 45 min to hard bake the photoresist S1813. This layer of S1813 acts as sacrificial layer in the subsequent processes to suspend the devices.
(2) The mark patterns that are used to precisely locate the coordinates of InAs nanowires are formed in the resist layer on Si/S1813 substrate by EBL, and they are converted into metal marks by depositing the metal film through EBE and following lift-off process.
(3) Mechanically transferring InAs nanowires from the growth substrate onto the substrate.
(4) A layer of electron-beam exposure resist PMMA 950 K (MicroChem) is spin-coated on the Si/S1813 substrate, and the designed source-drain electrode pattern is transferred on this layer of resist by EBL system.
(5) Using the $(NH_4)_2S_x$ solution to remove the native oxide layer on the surface of the InAs nanowires, in order to improve the contact between nanowires and electrodes.
(6) A Cr/Au/Cr metal film is deposited by EBE system, and the excessive deposited metal and electron-beam exposure resist are removed by a lift-off process.
(7) Conducting three-terminal electrical measurements on the fabricated back-gated FETs. The cured S1813 photoresist acts as gate dielectric of the back gate.

(8) Using the oxygen plasma to remove the sacrificial S1813 layer under the channel of FETs, so that makes the InAs nanowires suspended.
(9) Electrical measurements on InAs nanowire suspended devices.

The back-gated InAs nanowire device fabricated through these processes will be used for studying both room-temperature and low-temperature electrical characteristics of InAs nanowires with different crystal orientation in Chap. 4, while the suspended devices are used in Chap. 4 to transfer the InAs nanowires from the channel to TEM microgrid through manipulation technique.

2.4 Measurement and Characterization of InAs Nanowire Devices

In this work, electrical measurements of InAs nanowire devices are performed using room-temperature or low-temperature DC measurements, and they typically include two parts: measurement of transfer curves (source-drain current I_{DS} versus gate voltage V_G curve) and measurement of output curves (I_{DS} versus source-drain voltage V_{DS}).

Taking the back-gated devices as example, as shown in Fig. 2.13, we apply the V_{DS} across the two terminals (source and drain metal electrodes) and meanwhile measure the I_{DS} by placing two ports of the 4200SCS semiconductor analyzer at the source and drain electrodes. These two ports are equipped with front-end amplifier and thus have the highest measurement accuracy of 10^{-15} A. To regulate and evaluate the carrier concentration in the channel and the barrier height of contact barriers, we also use another port of Keithley 4200SCS semiconductor analyzer to apply a gate voltage on the heavily-doped silicon substrate which functions as a back gate. It should be noted that in order to reduce the thermal resistance and resistance between the silicon substrate and the measuring stage during low-temperature measurement, it is necessary to re-deposit a layer of metal (Ti/Au or Cr/Au) at the bottom of the silicon. The measurements involved in this work are all DC measurements. The scanning speed of V_{DS} used for the measurement is typically from 10 to 25 mV/s, that of V_{BG} from 1 to 2 V/s, and that of V_{TG} from 100 to 200 mV/s.

Since the conductance of a single InAs nanowires is generally low, the I_{DS} under the room-temperature measurement is on the order of 10^{-4} to 10^{-15} A, and under the low-temperature environment, I_{DS} can decrease to lower current level. Therefore, we need the measuring equipment and measuring lines that shield the external electromagnetic interference excellently. At the same time, in order to eliminate the impact of external environment, the measurement needs to be carried out in a vacuum environment. In the room-temperature measurement, the vacuum of the measurement environment is usually less than 1×10^{-3} mTorr. As to the low-temperature measurements, they require a higher vacuum environment which shall reach the order of 10^{-5} mTorr.

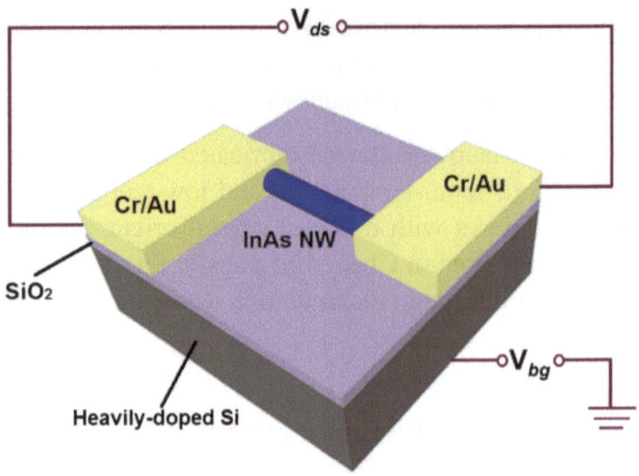

Fig. 2.13 Schematic diagram of device structure of InAs nanowire back-gated FET and their measurement circuit

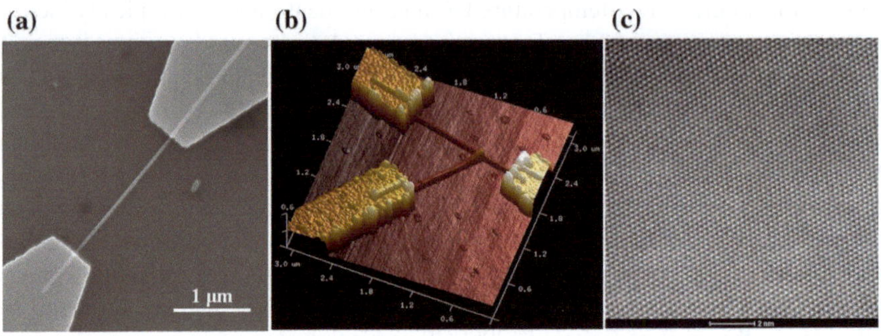

Fig. 2.14 **a** SEM characterization of a typical InAs nanowire back-gated FET; **b** AFM characterization of InAs nanowire FETs; **c** TEM characterization of InAs nanowires with single-crystal WZ phase structure. The scale bar in (**a**) is 2 nm

After completing the measurement of devices, we will characterize the morphology and structure of nanowire-based device, the crystal structure of nanowires and some other information, in order to determine the geometric parameters (including channel length, nanowires' diameter, etc.), crystal quality, crystal phase, growth direction and other factors that may affect the performance of nanowire-based devices. In the experiment, we usually use SEM to determine the channel length and structure of devices as shown in Fig. 2.14a, use AFM to determine the diameter of InAs nanowires and the surface topography of back-gated devices (Fig. 2.14b), and use TEM to obtain the crystal structure at atomic scale (Fig. 2.14c).

2.5 Extraction of Basic Electrical Parameters of InAs Nanowires Based on FETs

An FET is one kind of device structure that generates a longitudinal electric field by a gate voltage, thereby controlling the number and type of carriers in the channel to regulate the current across the channel. The FET is typically a three-terminal structure, which includes: one source, one drain, and one gate. As shown in Fig. 2.15, in conventional FETs fabricated by bulk materials, the contact and channel region is typically doped with different types of dopants, so that p-n junctions are formed between contacts and channel. These p-n junctions block the leakage current through the FETs at the OFF-state and consequently decrease the I_{OFF}. However, in this work, we use the nanowires to obtain a good pinch-off feature, because the one-dimensional nanowires have the advantage that their thickness of the channel in the axial direction is smaller than or comparable to the width of depletion region of the semiconductor material, as shown in Fig. 2.16, and thus the majority carriers in the channel can be almost completely depleted by the gate voltage [12]. Since the contact and channel portions of nanowires are uniformly doped (or intrinsic), the devices based on such nanowires are referred to as junctionless FETs or junctionless devices [12].

In Fig. 2.17b, we make a comparison between the transfer characteristics of Si nanowire junctionless devices and that of conventional trigate FETs. And based on this comparison, we can observe that nanowire-based junctionless FETs show similar current-voltage relations with p-n junction type FETs. Some basic electrical parameters of the InAs nanowire FETs, including the threshold voltage, ON-OFF ratio, and subthreshold swing, can be obtained by using the transfer characteristics and output characteristics of the FETs. Combined with the characteristics of the junctionless devices, some other basic electrical parameters of InAs nanowires, such as the carrier mobility, conductivity, and carrier concentration, can also be derived.

When the gate voltage is higher than threshold voltage, the nanowire junctionless FET has similar current conduction mechanism with conventional FET: drift current. When a small bias V_{DS} is applied to the drain, the I_{DS} increases linearly with V_{DS}, exhibiting similar characteristics with commonly-used resistors. Typically, under small V_{DS}, the total resistance of device is defined as the resistance in the linear region, $R = V_{DS}/I_{DS}$, and it comprises the contact resistance (R_{cont}), the channel resistance (R_{chan}) and other parasitic resistances (R_{acc}). If the channel length of the device is much larger than the mean free path of majority carrier and the carrier transport is in the diffusion region, the ideal current of device is $I_{DS} = \frac{Q_c}{\tau}$ [14], $\tau = \frac{L_g}{v} = \frac{L_g}{\mu \times E_{DS}}$, and $Q_c = C_g \times \left(V_G - V_T - \frac{V_{DS}}{2}\right)$. Here, Q_c is the electrical charge of majority carriers in nanowires, τ is the average transmission time of the majority carrier from source to drain, L_g is the channel length of device, v is the velocity of majority carrier, μ is the mobility of majority carrier, E_{DS} is the transverse electric field strength and C_g is the gate capacitance. Therefore we have [14]

$$I_{DS} = \frac{C_g \times \left(C_G - V_T - \frac{V_{DS}}{2}\right) \times \mu \times V_{DS}}{L^2} \tag{2.1}$$

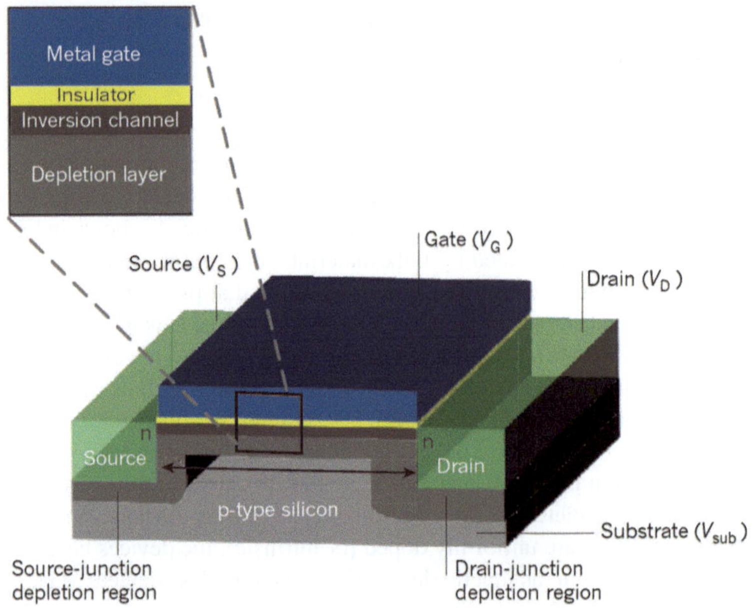

Fig. 2.15 Schematic diagram of a conventional n-type planar MOSFET. In the conventional n-type planar MOSFET, the bulk Si and the channel are typically doped with p-type dopants, and both the source and drain regions are doped with n-type dopants. When the gate voltage is not applied, the electrons are blocked by the p-n junctions formed between the contacts and channel, and thus the current across the devices is small; when a positive voltage is applied to the gate, an electron accumulation layer is formed at the surface of the channel to conduct the electrical current. Reprinted from Ref. [13], with kind permission from Springer Nature

When the device is at ON-state under small V_{DS}, I_{DS} and V_G satisfy a linear relation. If we extend the linear relation to intersect the x-axis (that is, $I_{DS} = 0$), then $V_G = V_T - V_{DS}/2$. In this work, the ON-state current (I_{ON}) denotes the source-drain current (I_{DS}) at the backgate voltage $V_{BG} = V_T + 10$ V and the OFF-state current (I_{OFF}) is the I_{DS} when I_{DS} in the transfer curve reaches the minimum at an negative enough V_{BG}. In this work, we can obtain V_T of devices through this mehod, and then define I_{DS} at $V_{BG} = V_T + 10$ V or $V_{TG} = V_T + 1$ V at the ON-state current I_{ON} of devices.

Taking the derivative of Eq. (2.1) to V_G, we get

$$\mu_{FE} = g_m L_g^2/(CV_{DS}) \tag{2.2}$$

The carrier mobility of nanowires can be obtained using Eq. (2.2). Here, g_m is the maximum value obtained by taking the derivative of I_{DS} and is called the maximum transconductance of devices. It should be noted that the carrier mobility extracted in experiments is often affected by contact resistance, gate parasitic capacitance and

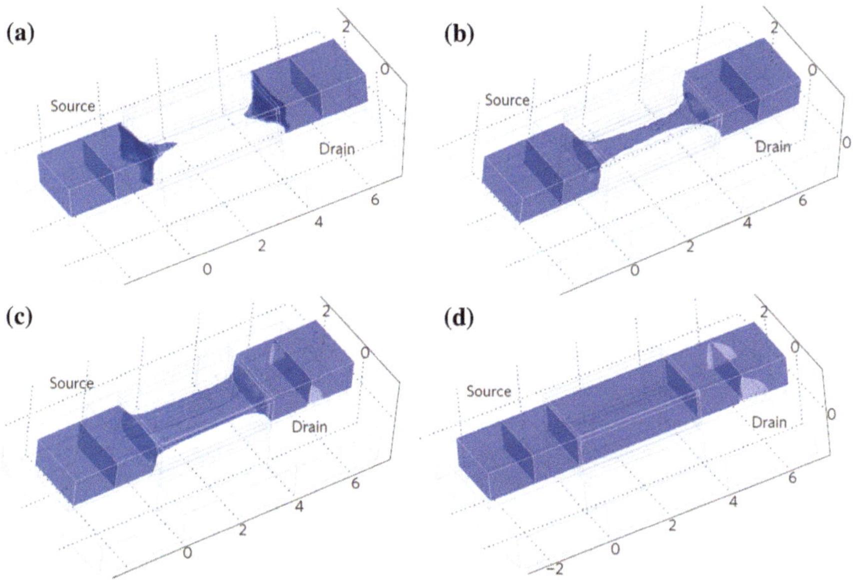

Fig. 2.16 Simulation of coutour lines of concentration of electrons in the body of n-type Si nanowire junctionless FETs at different gate voltages. The Si nanowire in the figure is doped with n-type dopants of 1×10^{19} cm^{-3}. The source-drain voltage of this device is 50 mV, its channel length is 20 nm, its thickness is 10 nm, and its width is 40 nm; **a** when the gate voltage is less than the threshold voltage, the nanowire device is at OFF-state, and the carriers in its channel are depleted; **b** when the gate voltage increases, drift electrons appear at the center of the nanowire and form electrical current; **c** when the gate voltage continues to increase, the region having high carrier concentration at the center becomes larger, and the electrical current continues to increase; **d** when the gate voltage reaches the flat band voltage, the electron concentration in Si nanowire is high in all the nanowire. Reprinted from Ref. [12], with kind permission from Springer Nature

non-ideal factors, so the extracted mobility can be smaller than the intrinsic carrier mobility of and it is generally referred to as the field-effect mobility.

Here, we extract μ_{FE} by using $\mu_{FE} = g_m L_g^2/(CV_{DS})$, where C_g is the back-gate capacitance to the nanowire. C_g is calculated by the metallic cylinder on an infinite metal plate model [15]:

$$\frac{C_g}{L_g} = 2\pi\varepsilon_0\varepsilon_r/\cos h^{-1}\left(\frac{t}{r}\right) \tag{2.3}$$

where ε_0 is the vacuum permittivity, ε_0, t, and r are the relative dielectric constant, the thickness of the dielectrics, and the radius of the InAs NWs, respectively.

Unlike conventional devices, the pinch-off of a junctionless device is achieved by depleting majority carriers. Therefore, the threshold voltage and some parameters associated with the OFF-state of devices are often related to the doping concentration, the thickness of the channel material and the effective carrier concentration at $V_G = 0$ V. In this work, since nanowires are undoped, we believe that the charge carriers

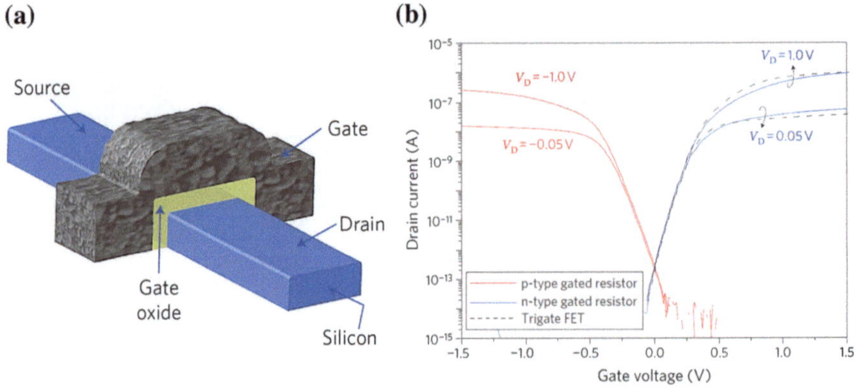

Fig. 2.17 **a** Schematic diagram of device structure of Si nanowire junctionless FETs. In the devices, the source and drain region have similar doping concentrations and doping types with the channel region. To completely deplete the majority carriers in nanowires and pinch off the nanowires, the thickness of nanowires are typically very small. **b** Comparison between the transfer characteristics of Si nanowire junctionless FETs and conventional Si-based Trigate FETs. The Si nanowire junctionless device has a channel length of 1 μm and width of 30 nm. Its transfer characteristics is similar to those of conventional Si-based devices Reprinted from Ref. [12], with kind permission from Springer Nature

are uniformly distributed along the nanowires when no gate voltage is applied. Since InAs nanowires conduct through electrons, if the contact resistance is not considered, the resistivity of InAs channel can be calculated by $\sigma = L_g/\pi r^2 R$ and the equivalent electron concentration n_e at the channel region can be calculated by $n_e = \sigma/e\mu_{FE}$ [16, 17]. When the majority carriers in nanowires are depleted, the current in devices no longer decreases, and the value of this current is I_{off} in this paper.

References

1. Krogstrup P, Jørgensen HI, Johnson E et al (2013) Advances in the theory of III–V nanowire growth dynamics. J Phys D Appl Phys 46(31):313001
2. de la Mata M, Zhou X, Furtmayr F et al (2013) A review of MBE grown 0D, 1D and 2D quantum structures in a nanowire. J Mat Chem C 1(28):4300
3. Fan HJ, Werner P, Zacharias M (2006) Semiconductor nanowires: from self-organization to patterned growth, small 2(6):700–717
4. Fortuna SA, Li X (2010) Metal-catalyzed semiconductor nanowires: a review on the control of growth directions. Semicond Sci Technol 25(2):024005
5. Zhang Y, Aagesen M, Holm JV et al (2013) GaAsP nanowires grown on silicon substrates by solid-source molecular beam epitaxy. Nano Letter 13(8):3897–3902
6. Messing ME, Hillerich K, Johansson J et al (2009) The use of gold for fabrication of nanowire structures. Gold Bull 42(3)
7. Thelander C, Caroff P, Plissard SB et al (2011) Effects of crystal phase mixing on the electrical properties of InAs nanowires. Nano Lett 11(6), 2424–2429

8. Joyce HJ, Gao Q, Hoe Tan H et al (2011) III–V semiconductor nanowires for optoelectronic device applications. Progr Quantum Electro 35(2–3):23–75
9. Dimakis E, Lähnemann J, Jahn U et al (2011) Self-assisted nucleation and vapor-solid growth of InAs nanowires on bare Si(111). Cryst Growth Des 11(9):4001–4008
10. Ghalamestani SG, Johansson S, Borg BM et al (2012) Uniform and position-controlled InAs nanowires on 2″ Si substrates for transistor applications. Nanotechnology 23(1):015302
11. Pan D, Fu M, Yu X et al (2014) Controlled synthesis of phase-pure InAs nanowires on Si(111) by diminishing the diameter to 10 nm. Nano Lett 14(3):1214–1220
12. Colinge J-P, Lee C-W, Afzalian A et al (2010) Nanowire transistors without junctions. Nat Nanotechnol 5(3):225–229
13. Ferain I, Colinge CA, Colinge JP (2011) Multigate transistors as the future of classical metal-oxide-semiconductor field-effect transistors. Nature 479(7373):310–316
14. Oktyabrsky S, Ye PD (2010) Fundamentals of III-V semiconductor MOSFETs. Springer, New York
15. Nilsson HA, Caroff P, Thelander C et al (2010) Temperature dependent properties of InSb and InAs nanowire field-effect transistors. Appl Phys Lett 96(15):153505
16. Dayeh SA, Aplin DP, Zhou X et al (2007) High electron mobility InAs nanowire field-effect transistors. Small 3(2):326–332
17. Ullah AR, Joyce HJ, Burke AM et al (2013) Electronic comparison of InAs wurtzite and zincblende phases using nanowire transistors. Physica Status Solidi—Rapid Res Lett **7**(10):911–914

Chapter 3
The Impact of Quantum Confinement Effects on Electrical Properties of InAs Nanowires

Abstract To suppress short channel effects, lower off-state leakage current and enhance gate coupling efficiency, InAs nanowires (NWs) with diameter smaller than 10 nm could be needed in field-effect transistors (FETs) as the channel length scales down to tens of nanometers to improve the performance and increase the integration. In this chapter, we report FETs based on ultrathin wurtzite-structured InAs NWs, with the smallest NW diameter being 7.2 nm. The FETs based on ultrathin NWs exhibit high Ion/Ioff ratios of up to 2 × 108, small subthreshold swings of down to 120 mV/decade, and operate in enhancement-mode. The performance of the devices changes as a function of the diameter of the InAs NWs. The advantage and challenge of the FETs based on ultrathin NWs are discussed.

Keywords InAs nanowires · Diameter-dependent · Electrical properties
Scaling down

InAs nanowires with diameter ranging from 7 to 16 nm, which are referred as "ultrathin nanowires", are used in this chapter as channel material of devices to systematically study variation of electrical parameters and device performance of InAs nanowires resulting from quantum confinement effects. To excluding the effect of other factors, ultrathin nanowires are all grown by MBE, so that the high-quality crystal structure and uniform growth direction can be obtained. So far, InAs nanowires studied in this chapter have the smallest diameter in the reported works.

3.1 Growth of Ultrathin InAs Nanowire

InAs nanowires studied in this chapter are grown on n-type Si (111) substrates by MBE through a two-step growth method [1]. Before moving the Si substrates into the MBE chamber, they were immerged in a diluted HF (2%) solution to remove the surface contamination and native oxides. A silver effusion cell was used for the catalyst deposition. As shown in Fig. 3.1a, the temperature of first step annealing is 200 °C and it lasts for 10 h. The temperature of the second annealing step is 600 °C

M. Fu, *Electrical Properties of Indium Arsenide Nanowires and Their Field-Effect Transistors*, Springer Theses, https://doi.org/10.1007/978-981-13-3444-3_3

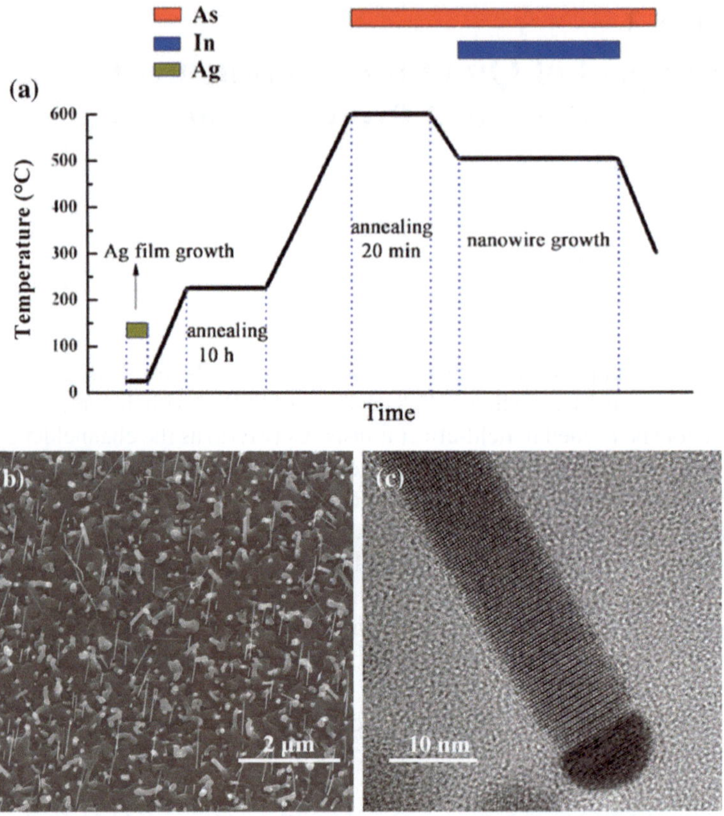

Fig. 3.1 **a** Temperature during growth of ultrathin InAs nanowires in MBE. Before growing nanowires, the two-step annealing method is used by applying a series of annealing temperature combined with different annealing time to obtain uniform Ag catalyst particles whose diameter are around 10 nm. **b** SEM images of ultrathin InAs nanowires on growth substrate. The ultrathin InAs nanowires are vertically grown on the growth substrate and the growth direction is <0001>. The sample is tilted by 30°. **c** High-resolution TEM image of the upper portion of one ultrathin InAs nanowire. The nanowire has no stacking faults and shows very high quality pure WZ phase. Reprinted with the permission from Ref. [1]. Copyright 2014 American Chemical Society

and it lasts for 20 min. After annealing, Ag forms relatively uniform Ag catalyst particles with diameters of 10 nm ± 3 nm.

InAs nanowires are nucleated at Ag catalyst particles and then grown perpendicular to substrate at 505 °C after introducing In and As gas sources. The ratio of As/In source is 30. As illustrated in Fig. 3.1c, The grown InAs nanowires show high-quality wurtize structure without stacking faults according to the TEM characterization results.

3.2 High-Performance Device Based on Ultrathin InAs Nanowires

In this part, the high-performance electrical devices are fabricated based on aforementioned high-quality wurtzite-phase InAs nanowires grown by MBE. These devices are top-gated with the fabrication processes introduced in Chap. 2. As illustrated in Fig. 3.1c, 25/50 nm Cr/Au and 5/90 nm Ti/Au were deposited as the source/drain (S/D) and top gate, respectively. A 12 nm thick HfO_2 layer was grown by ALD at 90 °C as the gate dielectric.

Figure 3.2c, d are the output and transfer characteristics of one typical device. The channel length of this device is 550 nm, and the diameter of the InAs NWs is 8.8 nm, which is determined by AFM and then deducts 1.5 nm-thick oxides. Notably, the subthreshold swing (SS) is 120 mV/decade for two decades and 65 mV/decade for one decade , and the I_{on}/I_{off} ratio is about 2×10^6. Here, I_{off} is the minimum

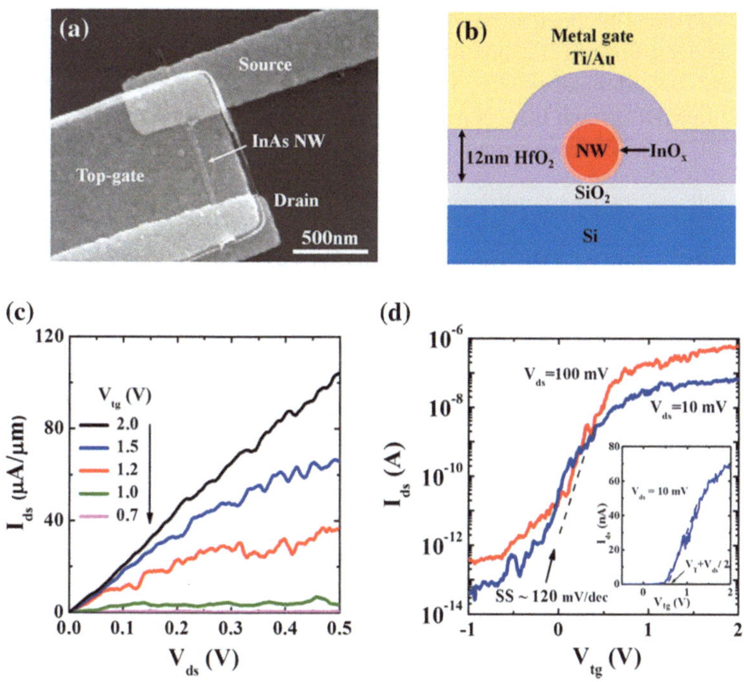

Fig. 3.2 **a** Top-view SEM image of a top-gate FET based on an ultrathin InAs nanowire. The nanowire has diameter of 8 nm, and the channel length of the device is about 550 nm. **b** Schematic diagram showing the cross-section of a top-gated InAs nanowire device. **c** Output characteristics (normalized to the nanowire's circumference) in linear scale. **d** Transfer characteristics in logarithmic scale. The insert graph is transfer characteristics in linear scale. Reprinted with permission from Copyright 2014, American Institute of Physics

Table 3.1 Comparison of subthreshold properties of reported InAs NW devices and ultrathin InAs nanowire FET (shown in Fig. 3.2)

D (nm)	L_G (nm)	SS (mV/dec)	I_{on}/I_{off}	Ref.	Technology
10	600	65 ($V_{ds} = 0.1$ V)	2×10^6	This work	Lateral NW
15	100	93 ($V_{ds} = 0.01$ V)	2×10^3	[2]	Lateral NW
55	750	98 ($V_{ds} = 0.5$ V)	1190	[3]	CMOS
25	170	80 ($V_{ds} = 0.01$ V)	10^3	[4]	Radial NW
50	50	88 ($V_{ds} = 0.5$ V)	3×10^4	[5]	Vertical NW

I_{ds} while I_{on} is the I_{ds} at $V_{tg} - V_T = 1$ V with V_{tg} being the top-gate voltage and V_T being the threshold voltage. This InAs FET also shows an enhancement-mode behavior with $V_T = 0.63$ V.

As summarized in Table 3.1, the present top-gated ultrathin InAs NWs device shows better subthreshold characteristics than previous lateral InAs nanowire-based devices and even the devices based on other technologies such as vertical gate-all-around devices. Specifically, our ultrathin InAs nanowire-based FETs show lower I_{off} by at least one order and smaller SS. Next, we evaluate the electrical characteristics of ultrathin MBE-grown InAs NWs with their diameter ranging from 7.2 to 16.1 nm through FETs and observe strong diameter-dependent performance in these devices.

3.3 Influences of Diameter on the InAs Nanowire Devices

We have studied more than 10 devices with nanowires' diameter of 7 to 16 nm. We find that by decreasing the diameter of the NWs, a higher I_{on}/I_{off} ratio, lower I_{off} and more positive V_T can be achieved, while the advantages of InAs NW devices on low R_{cont}, high μ_{FE} and high ON-state resistance (R_{on}) are compromised.

3.3.1 *Influences of Diameter on the OFF-State Performance of InAs Nanowire Devices*

As shown in Fig. 3.3a, when the diameter of InAs nanowires increases from 12 to 16 nm, the ON-OFF ratio of top-gated devices is about 10^4, and does not show strong dependence on the diameter. However, when the diameter of the InAs NWs is smaller than 12 nm, the I_{on}/I_{off} ratio increases linearly in a logarithmic coordinates with decreasing diameter. As shown in Fig. 3.4, the maximum I_{on}/I_{off} ratio we obtained in a device based on an InAs NW with 7.2 nm diameter is 10^8 and I_{off} is smaller than the limit of measurement accuracy (fA) of the measurement system. In Fig. 3.3b, the V_T shifts to more positive value as the diameter of the NWs decreases, which shows

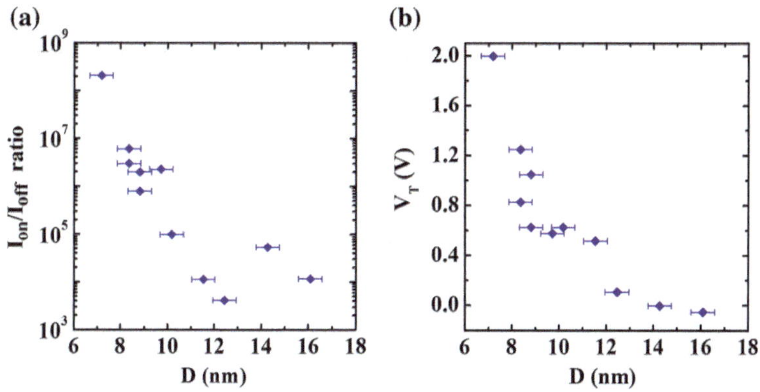

Fig. 3.3 **a** The I_{on}/I_{off} ratio and **b** V_T measured from top-gated FETs based on InAs NWs having different diameters. The I_{on}/I_{off} ratio and V_T change with the diameter of the NWs. Reprinted with permission from Copyright 2014, American Institute of Physics

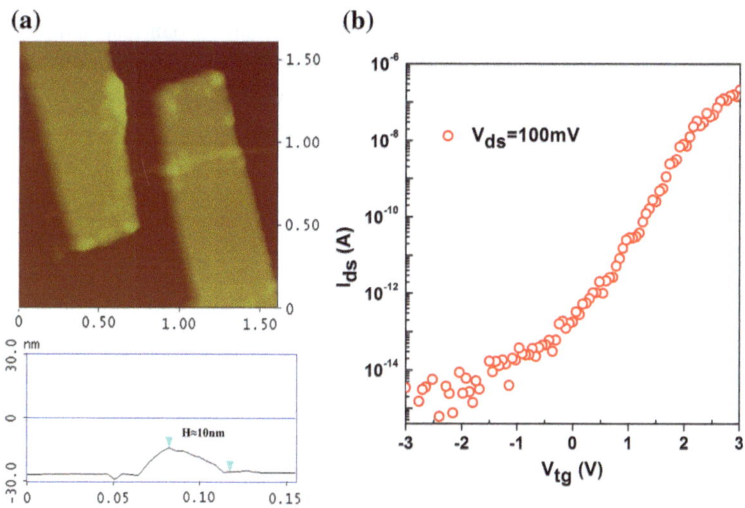

Fig. 3.4 **a** AFM image and line profile along the dotted line taken from InAs nanowire having 7 nm diameter after deducting 1.5 nm-thick oxides. **b** Transfer characteristic of top-gated device based on this 7 nm InAs nanowire

similar dependence on diameter with $I_{\mathrm{on}}/I_{\mathrm{off}}$ ratio. When the diameter is larger than 12 nm, V_T is just above zero. And when the diameter is smaller than 12 nm, V_T shifts to positive voltage and the devices all operate in enhancement-mode. When the diameter is smaller than 10 nm, V_T changes about 0.2 V/nm. This diameter dependence is unfavorable for FETs.

Due to the narrow bandgap ($E_{g,bulk} = 0.36$ eV) and high density surface states [6] near the conduction band, devices based on bare InAs NWs in previous reports do not have satisfactory performance in $I_{\mathrm{on}}/I_{\mathrm{off}}$ ratio and SS and present depletion-mode characteristics even with a very thin high-k dielectric layer [7]. In this chapter, we use ultrathin InAs nanowires as channel material, which enhances the strong electrostatic gate coupling between the channel materials and the top-gate, so that the switching characteristic of the devices is largely enhanced [8]. This factor can account for the slow change of the subthreshold properties when the diameter of NWs is larger than 12 nm. It is significant to further scale down the devices.

3.3.2 Larger Bandgap Induced by Smaller Diameter of InAs Nanowires

Zhao et al. reported that, the I_{on}/I_{off} ratio of FETs based on one-dimensional narrow-bandgap material is related to their bandgap, and their relation can be approximated by a simple model [9]. By comparing their model with reported works of InAs nanowires [4], InSb nanowires [10–12] and carbon nanotubes [13], Zhao's model shows its rationality and universality in terms of the narrow bandgap semiconductor materials. Accordingly, we use this model to estimate the bandgap of ultrathin InAs nanowires in this chapter. Accordig to n-type performance of InAs nanowire-based FETs, when V_{DS} is small, we use the following equation to estimate the bandgap of InAs nanowires:

$$\frac{I_{on}}{I_{off}} \approx \frac{1}{4}\exp\left(\frac{E_g - \Delta E}{k_B T}\right) \tag{3.1}$$

Here, k_B is the Boltzmann constant, T is the temperature, and ΔE is the effective barrier reduction value introduced by the contact barrier and the tunneling current. If E_g-ΔE in Eq. (3.1) is defined as the effective barrier height in OFF-state, Eq. (3.1) can be simplified into

$$\frac{I_{on}}{I_{off}} \approx \frac{1}{4}\exp\left(\frac{E_{SB}^*}{k_B T}\right) \tag{3.2}$$

Ideally, the maximum I_{on}/I_{off} ratio achieved by InAs nanowires FETs with a bandgap of 0.36 eV is approximately 10^5, and it is consistent with the reported experimental works. In this chapter, the crystal structure of InAs nanowires is WZ, and according to previous theoretical and experimental reports, the bandgap of InAs material with WZ phase is larger than the ZB phase by tens of meV (Fig. 3.5).

With a large Bohr radius about 34 nm, InAs nanowire is expected to have larger bandgap, sub-band separation, and reduced density of states when the diameter is decreased [14–16]. Since there is no directly relevant study on the bandgap of WZ phase nanowires with diameter from 7 to 16 nm, we will qualitatively study the performance of WZ-phase ultrathin InAs nanowire FETs referring to the relation between the bandgap and the diameter in ZB-phase InAs nanowires.

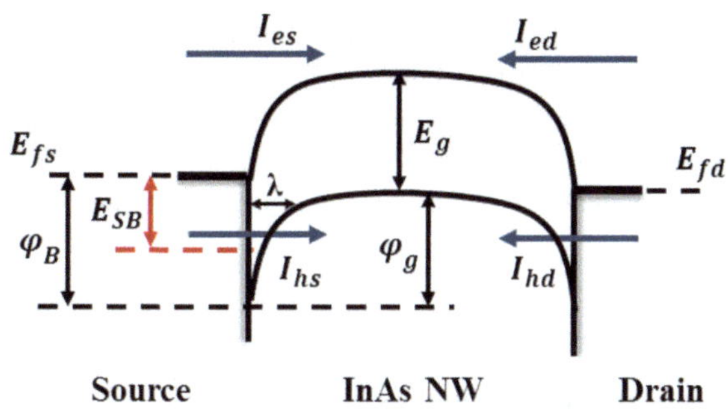

Fig. 3.5 The schematic diagram shows the band structure of InAs NW-based FET at OFF-states. In this figure, E_{fs} and E_{fd} are the Fermi level at source and drain, respectively. The E_{SB} is the height of effective barrier that blocks electrons from the source into InAs nanowires. The height of the effective barrier is related to the effective mass of the carrier, the height (φ_B) and effective width (λ) of the barrier between contact interface and nanowire channel, and so on. The total current in InAs NW device is the sum of four components: source injection electron current I_{es}, drain injection electron current I_{ed}, source tunneling hole current I_{hs}, and drain tunneling hole current I_{hd}. In the ideal junctionless InAs nanowire FET, the electron injection current $I_{es} + I_{ed}$ is equal to the hole tunneling current $I_{hs} + I_{hd}$ at OFF-state. At this time, since the Fermi level of the region of InAs nanowire that is at the interface between metal contact and nanowire are pinned above the edge of conduction band of InAs nanowires, the effective barrier height is similar to E_g

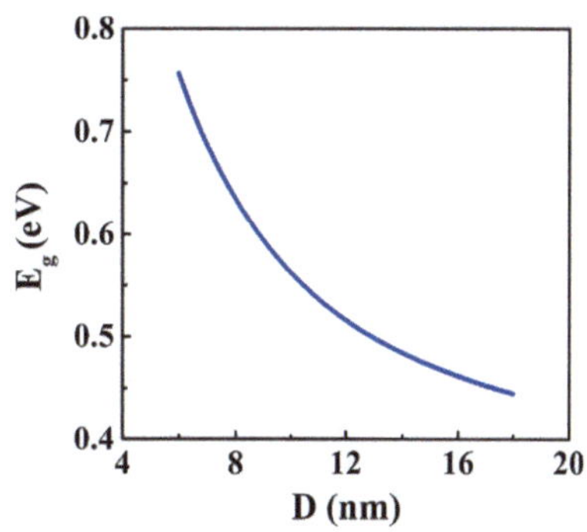

Fig. 3.6 The relation between theoretically calculated bandgap and diameter of ZB-phase InAs nanowire

Wang et al. reported experimental value of bandgap of ultrathin ZB-phase InAs nanowires measured by photoluminescence spectroscopy [16]. According to Wang's work, the experiential results on ultrathin ZB-phase InAs nanowires is consistent with the theoretical calculation, when diameter of InAs nanowires ranges from 5 to 14 nm. According to their theoretical calculation, as shown in Fig. 3.6, the relation between diameter (or radius) and the bandgap of ZB-structure InAs nanowires follows the equation [14]

$$E_g(R) = \frac{10.6951}{R^2 + 6.512R + 2.773} + \frac{0.9887}{R^2 + 2.720R + 0.898} + E_g(\infty) \tag{3.3}$$

Comparing Fig. 3.6 with Fig. 3.3, we find that the relation between the bandgap and the diameter is similar to that between $\log(I_{on}/I_{off})$ and the diameter, as well as the relation between V_T and diameter. And this similarity can also further prove that the higher I_{on}/I_{off} ratio and more positive V_T in thinner InAs nanowire FETs is attributed to the larger bandgap caused by quantum confinement effects.

3.3.3 Influence of Diameter on ON-State Performance of InAs Nanowire Devices

Though scaling down of InAs nanowires brings improvement on the subthreshold behavior of their FETs, the field-effect mobility of carriers, in particular electrons, in ultrathin InAs nanowires is reduced as the diameter becomes smaller due to the more severe surface roughness scattering to electrons as a result of a large surface to volume ratio. As shown in Fig. 3.7, when the channel length is between 150 and 250 nm, the electron μ_{FE} of back-gated nanowire devices show a linear correlation with diameter of nanowires.

The lower μ_{FE} leads to smaller ON-state current. As shown in Fig. 3.7a, the ON-state resistance (R_{on}) of InAs nanowire with a diameter of 7 nm increases by one magnitude comparing with that of 10 nm. However, though R_{on} still decreases when the diameter of InAs nanowire is above 10 nm, the difference between InAs nanowires with various diameter becomes relatively small. As shown in Fig. 3.7a, the R_{on} of InAs nanowire with a diameter of 10 nm is twice of that of 14 nm.

The ON-state current can be improved by scaling down the L_g. As shown in Fig. 3.8, we fabricated multiple Cr/Au electrodes to form several FETs on the same

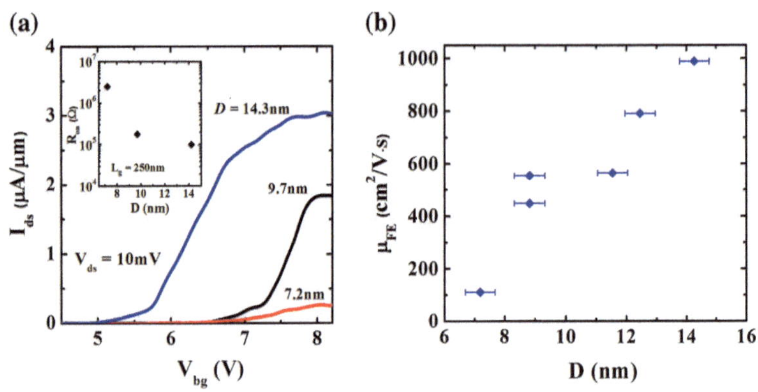

Fig. 3.7 **a** Tranfer characteristics of back-gated InAs nanowire FETs with various diameter, wherein V_{DS} = 10 mV, L_g = 250 nm. The inserted graph shows relation between extracted R_{on} and diameter of InAs nanowires. **b** Relation of μ_{FE} of back-gated nanowire FETs and diameter of InAs nanowires, where the L_g of studied nanowire devices ranges from 150 to 250 nm. Reprinted with permission from Copyright 2014, American Institute of Physics

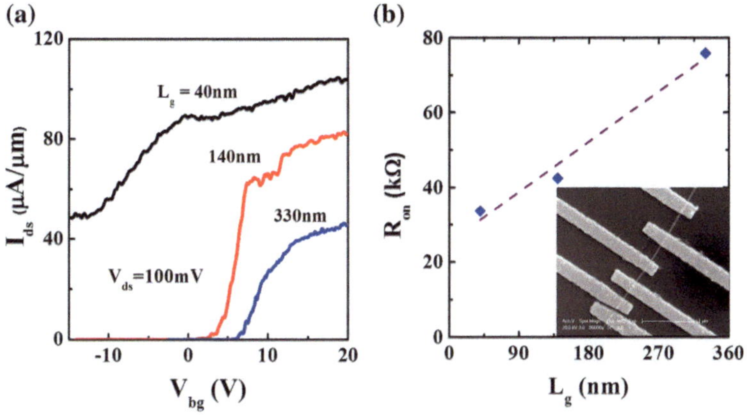

Fig. 3.8 **a** Transfer characteristics of back-gated InAs nanowire FETs fabricated from the same ultrathin InAs nanowires. This ultrathin InAs nanowire has diameter around 9 nm and channel length from 40 to 330 nm. **b** R_{on} versus channel length of FETs. The inserted graph is SEM image of the back-gated InAs nanowire FETs having multiple contact electrodes

InAs nanowire with diameter about 9 nm. The channel length of these FETs varies from 40 to 330 nm. The R_{on} linearly decreases with L_g. when L_g is 330 nm, R_{on} of FET is about 76 kΩ. And when L_g is 40 nm, R_{on} decreases to 40 kΩ. The linear relation between R_{on} and L_g shows that when L_g is 40 nm, the device still works in diffusive region, so that the mean free length of the nanowire is supposed to be smaller than 40 nm. With the transmission line method (TLM) [17], we extend the curve of R_{on} to $L_g = 0$, and extract the contact resistanc to be around 25 kΩ in total. Therefore, when the channel length of nanowire-based FET is short the contact resistance between ultrathin nanowires and metal electrodes contribute a major portion to the total resistance of the device.

If we ignore other parasitic resistances, when V_{DS} is relatively small, the electrical properties of InAs nanowire devices satisfy the following equation

$$V_{DS} = I_{DS} \times (2 \times R_{cont} + R_{channel}) = \left(R_{cont} \times I_{DS} + V_{DS}^{*}\right) \tag{3.4}$$

Here, V_{DS} is the working voltage applied to the source and drain electrodes of devices, and V_{DS}^{*} is the actual voltage drop across the channel part of devices. According to the above equation, we know that when applying a V_{DS} across the source/drain electrodes, the contact resistance acts as a voltage divider on the V_{DS}, which reduces the actual voltage drop that is used to accelerate carriers, and results in lower current through the device and slower working speed. From the equation of extracting μ_{FE}:

$$\mu_{FE} = g_m L_g^2 / \left(C_g V_{DS}\right) \tag{3.5}$$

we know that the influence of contact resistance results in a smaller mobility than the intrinsic value. We compare the contact resistance of nanowires with different

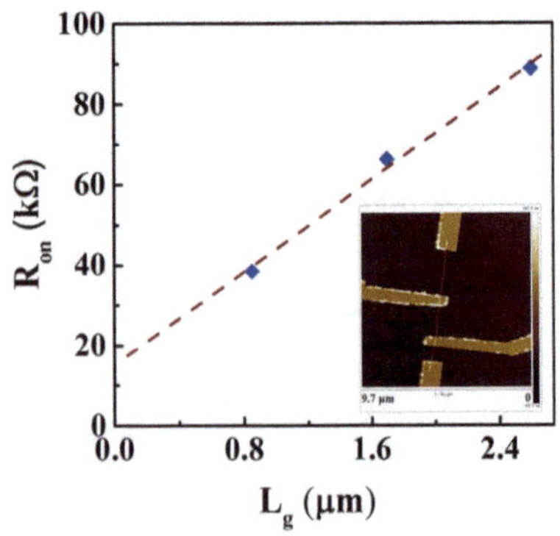

Fig. 3.9 ON-resistance of devices based on a same InAs nanowire with different channel length. The inset is an AFM image of the device. The diameter of this InAs nanowire is approximately 23 nm

diameter by TLM method shown in Figs. 3.8 and 3.9, and find that the contact resistance decreases with the increasing diameters. Consequently, for the application of ultrathin InAs nanowires, it is a critical to reduce the contact resistance of the devices.

3.4 Summary

In this chapter, we have fabricated and studied FETs based on MBE-grown WZ InAs NWs thinner than 10 nm. The FETs show favorable subthreshold characteristics and enhancement-mode operation because of the quantum confinement effects as well as the strong electrostatic coupling between channel and gate in ultrathin InAs nanowire-based devices. Specifically, the I_{on}/I_{off} ratio increases drastically with decreasing the diameter and can reach 2×10^8 when the diameter is 7.2 nm. This is the highest I_{on}/I_{off} ratio in reported works. The threshold voltage also increases as the diameter decreases, and enhancement-mode behavior is observed in all the devices when the diameter of the NW is smaller than 14 nm. On the other hand, the decreasing of ON-state current with decreased diameter is a great challenge to fabricate high speed FETs using the ultrathin InAs NWs in the future. The field-effect mobility of the WZ-structured InAs NW decreases linearly with the diameter of the NW as a result of strong surface scattering. The contact resistance is also a major concern for ultrathin NWs.

References

1. Pan D, Fu M, Yu X et al (2014) Controlled synthesis of phase-pure InAs nanowires on Si(111) by diminishing the diameter to 10 nm. Nano Lett 14(3):1214–1220
2. Dey AW, Thelander C, Lind E et al (2012) High-performance InAs nanowire MOSFETs. IEEE Electron Device Lett 33(6):791–793

3. Dey AW, Svensson J, Borg BM et al (2012) Single InAs/GaSb nanowire low-power CMOS inverter. Nano Lett 12(11):5593–5597
4. Jiang X, Xiong Q, Nam S et al (2007) InAs/InP radial nanowire heterostructures as high electron mobility devices. Nano Lett 7(10):3214–3218
5. Thelander C, FrobergFroberg LE, Rehnstedt C et al (2008) Vertical enhancement-mode InAs nanowire field-effect transistor with 50-nm wrap gate. IEEE Electron Device Lett 29(3):206–208
6. Hjort M, Lehmann S, Knutsson J et al (2014) Electronic and structural differences between wurtzite and zinc blende InAs Nanowire surfaces: experiment and theory. ACS Nano 8(12):12346–12355
7. Dayeh SA, Soci C, Yu PKL et al (2007) Transport properties of InAs nanowire field effect transistors: the effects of surface states. J Vacuum Sci Technol B: Microelectron Nanometer Struct 25(4):1432
8. Takei K, Chuang S, Fang H et al (2011) Benchmarking the performance of ultrathin body InAs-on-insulator transistors as a function of body thickness. Appl Phys Lett 99(10):103507
9. Zhao Y, Candebat D, Delker C et al (2012) Understanding the impact of Schottky barriers on the performance of narrow bandgap nanowire field effect transistors. Nano Lett 12(10):5331–5336
10. Nilsson HA, Caroff P, Thelander C et al (2010) Temperature dependent properties of InSb and InAs nanowire field-effect transistors. Appl Phys Lett 96(15):153505
11. Das SR, Delker CJ, Zakharov D et al (2011) Room temperature device performance of electrodeposited InSb nanowire field effect transistors. Appl Phys Lett 98(24):243504
12. Paul RK, Penchev M, Zhong J et al (2010) Chemical vapor deposition and electrical characterization of sub-10 nm diameter InSb nanowires and field-effect transistors. Mater Chem Phys 121(3):397–401
13. Javey A, Guo J, Farmer DB et al (2004) Self-aligned ballistic molecular transistors and electrically parallel nanotube arrays. Nano Lett 4(7):1319–1322
14. Niquet Y, Lherbier A, Quang N et al (2006) Electronic structure of semiconductor nanowires. Phys Rev B 73(16)
15. Ford AC, Kumar SB, Kapadia R et al (2012) Observation of degenerate one-dimensional subbands in cylindrical InAs nanowires. Nano Lett 12(3):1340–1343
16. Wang F, Yu H, Jeong S et al (2008) The scaling of the effective band gaps in indium–arsenide quantum dots and wires. ACS Nano 2(9):1903–1913
17. Mohney SE, Wang Y, Cabassi MA et al (2005) Measuring the specific contact resistance of contacts to semiconductor nanowires. Solid State Electron 49(2):227–232

Chapter 4
Influence of Crystal Phase and Orientation on Electrical Properties of InAs Nanowires

Abstract In this chapter, a systematic study on the correlation of the electrical properties with the crystal phase and orientation of single-crystal InAs nanowires (NWs) grown by molecular-beam epitaxy. A new method is developed to allow the same InAs NW to be used for both the electrical measurements and transmission electron microscopy characterization. We find both the crystal phase, wurtzite (WZ) or zinc-blende (ZB), and the orientation of the InAs NWs remarkably affect the electronic properties of the field effect transistors based on these NWs, such as the threshold voltage (VT), ON-OFF ratio, subthreshold swing (SS) and effective barrier height at the off-state (Φ_{OFF}). The SS increases while VT, ON-OFF ratio and Φ_{OFF} decrease one by one in the sequence of WZ <0001>, ZB <131>, ZB <332>, ZB <121> and ZB <011>. The WZ InAs NWs have obvious smaller field-effect mobility, conductivities and electron concentration at VBG = 0 V than the ZB InAs NWs, these parameters are not sensitive to the orientation of the ZB InAs NWs. We also find the diameter ranging from 12 to 33 nm shows much less effect than the crystal phase and orientation on the electrical properties of the InAs NWs. The good ohmic contact between InAs NWs and metal remains regardless of the variation of the crystal phase and orientation through temperature dependent measurements. Our work promotes deeper understanding of InAs NWs and is important for the development of nanowire-based devices.

Keywords InAs nanowires · Electrical properties · Crystal phase
Crystal orientation · Nano-manipulation

According to previous theoretical and experimental works, ZB-phase InAs material has a smaller effective mass than WZ-phase InAs material [1–7]. Moreover, InAs nanowires grown in different orientations can have different bandstructures even with the same crystal structures, and can be influenced by quantum confinement effects [8–10]. As a result, the research on the electrical properties of InAs nanowires with different crystal phases and crystal orientation can provide new ideas for optimizing InAs nanowire devices. For instance, the ZB-phase InAs nanowires are advanta-

M. Fu, *Electrical Properties of Indium Arsenide Nanowires and Their Field-Effect Transistors*, Springer Theses, https://doi.org/10.1007/978-981-13-3444-3_4

geously used to improve ON-state performance. However, the previous works on InAs nanowires cannot provide a clear relation among crystal phase and electrical properties of InAs nanowires, because the influence of the crystal phase is often accompanied with disorders like stacking faults, mixed phases [3, 5], background carbon doping [11, 12], and even geometries [3, 6]. In particular, in previous works, the crystal phase of nanowires in devices can not be directly characterized, consequently leading to the unclarity of the impact of crystal structure of InAs nanowires. To obtain a clear understanding on this issue, it is desired to choose a clean and simple material system (without doping, similar in size). As to the impact of crystal orientation on electrical properties of InAs nanowires, almost no relevant experimental research has been reported.

In this chapter, we first develop a new method to establish the connection between the crystalline structure and the physical properties of the same individual InAs NWs by combining nano-manipulation and nanolithography techniques. We can achieve for the first time an explicit correlation between the electrical transport properties and the crystalline orientation as well as the crystal phase of individual InAs NWs. We then quantitatively study the electrical transport properties of InAs NWs with different phase and orientation.

4.1 Growth of InAs Nanowires in Different Orientations

The InAs nanowires used in this chapter are epitaxially grown by MBE [13]. The substrates used in growth is similar with these in Chap. 3. A 2-inch substrate precleaned with hydrofluoric acid (HF) was placed in the MBE chamber. After the in situ deposition of 0.2 nm Ag film in the chamber, a layer of catalyst was formed by a conventional one-step annealing method. The InAs nanowires were grown after annealing at 600 °C for 20 min. The growth temperature was then set to 505 °C and the V/III source ratio was 30.

By characterizing structure and geometry of obtained nanowires, we found that diameter of InAs nanowires is quite uniform over the entire 2-in. substrate. However, the orientations of these nanowires are very different. Specifically, even in one small area, we can observe by SEM many nanowires that are vertically grown as well as grown at a angle with substrate. From TEM characterization, we found that most of the InAs nanowires show high-quality crystal structure but have different crystal phases (WZ or ZB phase). Moreover, the composition of nanowires with different crystal structures was characterized by EDS in TEM (shown in Table 4.1) and all show atomic ratio of Into As is very close to 1:1 (Figs. 4.1 and 4.2).

Table 4.1 EDS spectrum analysis results of multiple InAs nanowires

Number of nanowires	Crystal phase	Ratio of atoms (%)	
		As	In
1	ZB	51.1	48.9
2	ZB	51.7	48.3
3	ZB	52.9	47.1
4	WZ	52.7	47.3
5	WZ	52.9	47.1
6	WZ	52.0	48.0
7	WZ	52.4	47.6
8	WZ	50.6	49.4
9	WZ	52.1	47.9

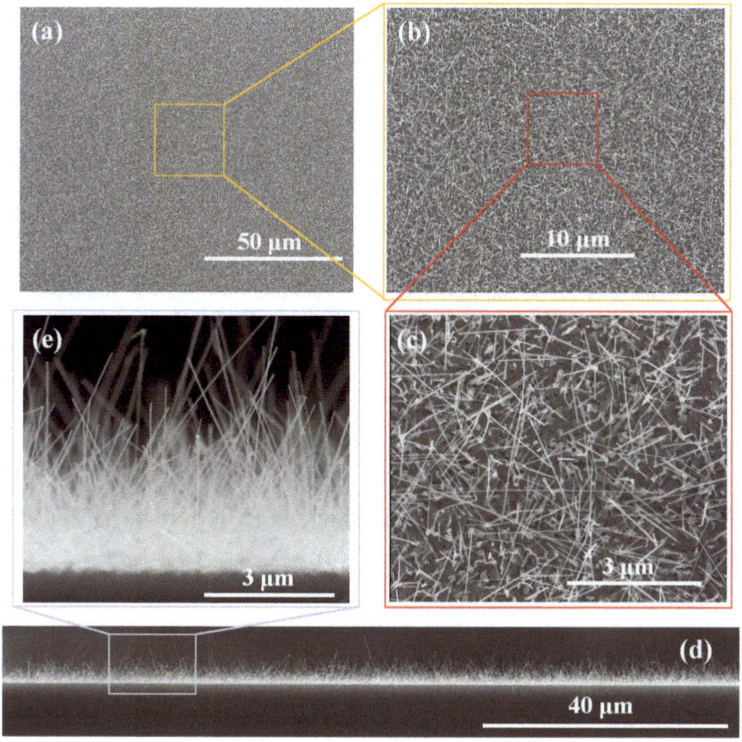

Fig. 4.1 **a–c** Top view of InAs nanowires growth substrate by SEM characterization. The magnification factor from (**a–c**) increases gradually. **d–e** Cross-section view of InAs nanowires growth substrate by SEM characterization

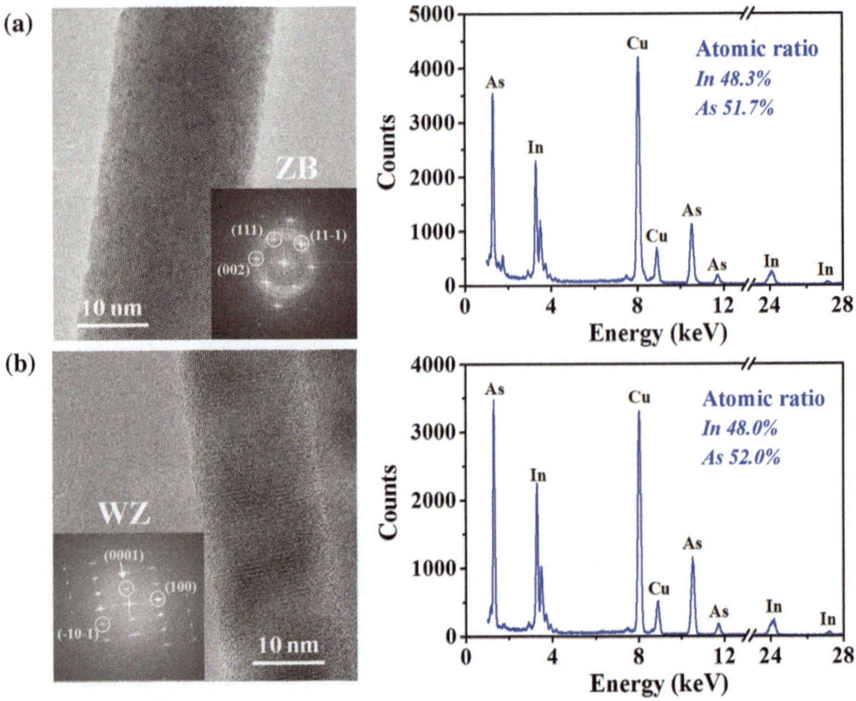

Fig. 4.2 TEM characterization and EDS spectrums of **a** one ZB InAs NW and **b** one WZ InAs NW. The insert graphs is FFT of corresponding TEM images. Both WZ-phase and ZB-phase InAs nanowires are single crystal with no or few stacking faults, and the ratio of In atoms to As atoms in these two kinds of InAs nanowires are all near 1:1. No difference on the stoichiometry is observed between the WZ and ZB InAs NWs. Reprinted with the permission from Copyright 2016 American Chemical Society

4.2 Method of Transferring Nanowires in Devices to TEM for Structural Characterization

In this chapter, to investigate the correlation between the structural and the electrical transport properties of the same individual InAs NWs, we develop the following experimental method and process, as shown schematically in Fig. 4.3. This method comprises three key steps:

(1) the photoresist (S1813) needs to be hard baked at 200 °C before the NW transfer so that it will not be dissolved by the organic solvents used in the following processes, and then the cured photoresist acts as dielectrics in unsuspended devices and a sacrificial layer for suspending the InAs nanowires;

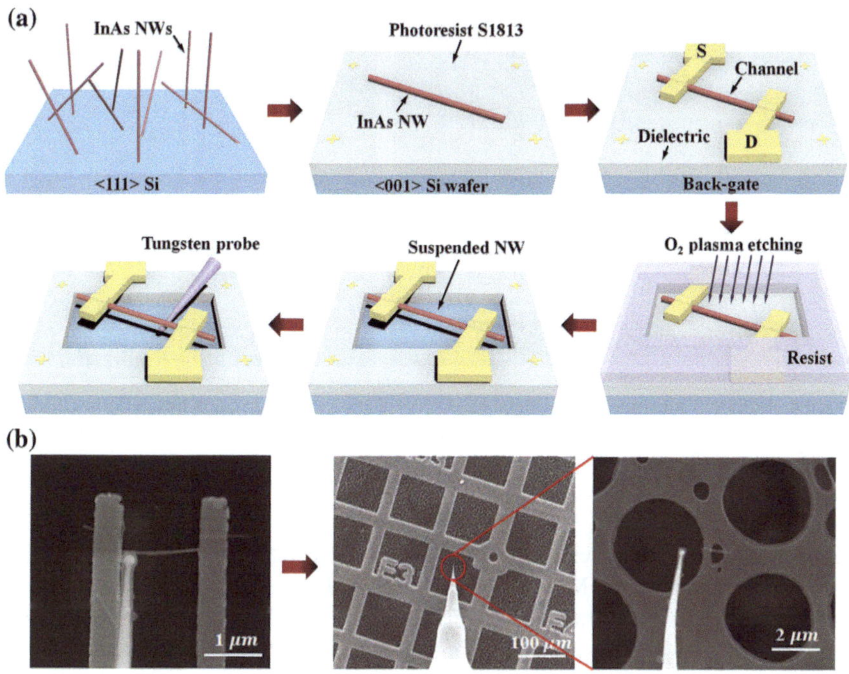

Fig. 4.3 **a** Schematic process for fabricating suspended InAs nanowire devices. **b** SEM images showing an InAs nanowire transferred by nano-manipulation from a FET to a holey carbon film supported by Cu grid. There are metal marks on the Cu grid for locating the nanowires in HRTEM characterization. After sticking the nanowire to the carbon film, SEM images in different magnifications are taken to locate the InAs nanowires. Reprinted with the permission from Copyright 2016 American Chemical Society

(2) after all the room temperature and low temperature electrical measurements are finished, the InAs NWs in the devices are suspended through etching the photoresist under the channel by O_2 plasma treatment, so that the InAs nanowires can be less damaged and the process for suspending nanowires is much simplified;
(3) through nano-manipulation in SEM, the suspended NWs are transferred [14, 15] onto a carbon film supported by Cu grids.

It should be noted that before the NWs are mechanically transferred to a n-doped Si substrate coated with a layer of 1.2 μm-thick S1813 photoresist, patterned marks have been fabricated in advance on the S1813 layer (as shown in Fig. 4.4) to precisely locate the dispersed NWs. Individual NWs with enough length (longer than 2.3 μm) and homogeneous diameter (generally having diameter of 22 nm ± 10 nm) along their growth axis are then located by SEM and selected for the device fabrication. To exclude possible effect of the channel length (L_g) to the electrical properties, all the FETs studied here are designed to have a same L_g of around 1.5 μm. This long

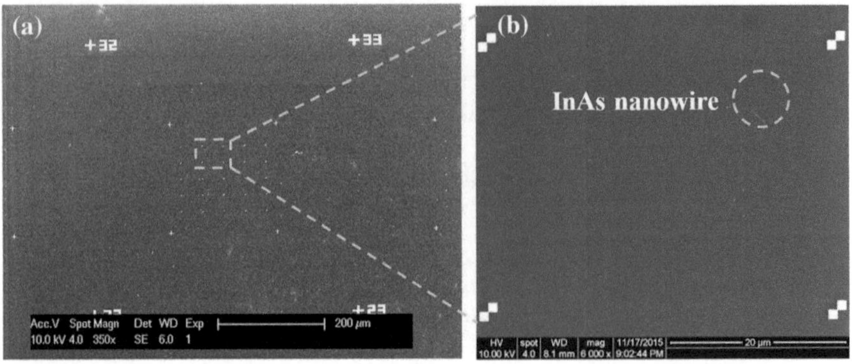

Fig. 4.4 Si/S1813 substrate having metal marks for locating the InAs nanowires. **a** Shows the large cross mark with Arabic numerals used for coarsely locating nanowires, and **b** Shows the small metal mark for accurately locating nanowires

channel length also ensures nanowire-based FETs to work in diffusive region with a low contact resistance. More importantly, this channel length is tested to be the optimized length for following manipulation processes. When the channel length is shorter than 1.5 μm, the suspended nanowires are too short to be picked up by the probes to the carbon holes on the microgrid. While when channel length is longer than 1.5 μm, the nanowires tend to stick to the device substrate possibly because of a large electrostatic force between nanowires and the substrate.

After all the room temperature and low temperature electrical measurements are finished, the InAs NWs in the devices are suspended through O_2 plasma etching. Then, through nano-manipulation in SEM, we first move the probe under the suspended nanowires, and contact with the nanowire and fix it to the nanowire by depositing amorphous carbon. After confirming that the nanowire is firmly anchored to the tip, we in situ break the ends of the wire from the edge of the metal electrodes, and then transfer nanowire to the upper space of a carbon film supported by Cu grids with marks. At the selected location, the nanowire on probe is slowly placed flat down on the carbon film, and amorphous carbon is deposited again by electron beam irradiation to connect the InAs nanowires and the carbon film. After firmly anchoring the nanowire to the carbon film, we shake the probe repeatedly to separate probe and nanowire. To ensure the correspondence between TEM characterization and electrical measurements, we place nanowires from different devices in different areas of the micro-grids that are labeled by marks. For example, such a "E3" label is shown in the middle image of Fig. 4.3b. Besides, a series SEM images are taken at different magnification with the nanowires on the carbon film are used to identify the position in TEM. The InAs nanowires transferred to the coordinate microgrids were characterized by HRTEM to determine their diameter and crystal quality, as well as their crystal phase and crystal orientation.

4.3 Determination of the Crystal Orientation

The crystal phase and orientation of the InAs nanowires are determined from the FFT of TEM images. In the following we introduce in detail two methods to determine the growth direction.

In the first case, if different parts of a targeted InAs nanowire are shown in almost the same focus plane in TEM, we consider the whole InAs nanowire is parallel to the focus plane of TEM which is vertical to incident direction of electron beam. In this case, we simply find the diffraction spot (marked as point C) whose connection with the transmission spot O is parallel to the direction of the nanowires. And the planar index of this diffraction spot is the same to the crystal orientation of the nanowire. Therefore, only one TEM high resolution photo is required to determine the growth direction.

However, if the InAs nanowires are not parallel to the focus plane or tilted relative to the incident direction, we use the second method for calibration. In this case shown in Fig. 4.5, two high-resolution TEM images of targeted InAs nanowire taken from different zone axis direction are needed. First, two diffraction spots (marked as point D1 and D2, respectively) whose connection with transmission spot O is parallel to the direction of the nanowires are found in FFTs of two different HRTEM images. The crystal orientation of this nanowire is parallel to the direction calculated by $\overrightarrow{OD1} \times \overrightarrow{OD2}$. As shown in Fig. 4.5, D1 in Fig. 4.5a is the diffraction spot corresponding to the (133) lattice plane, and D2 in Fig. 4.5b is the diffraction spot corresponding to the (011) lattice plane. In this way, we can determine the growth direction of this nanowire to be along the direction of $(133) \times (011)$.

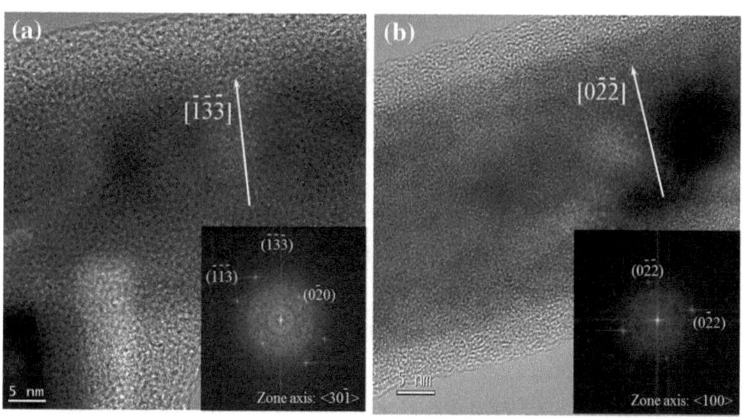

Fig. 4.5 HRTEM images taken from a same InAs nanowire, when the electron beam is incident from different zone axis directions. **a** The incident direction is parallel to the <301> direction; **b** The incident direction is parallel to the <100> direction

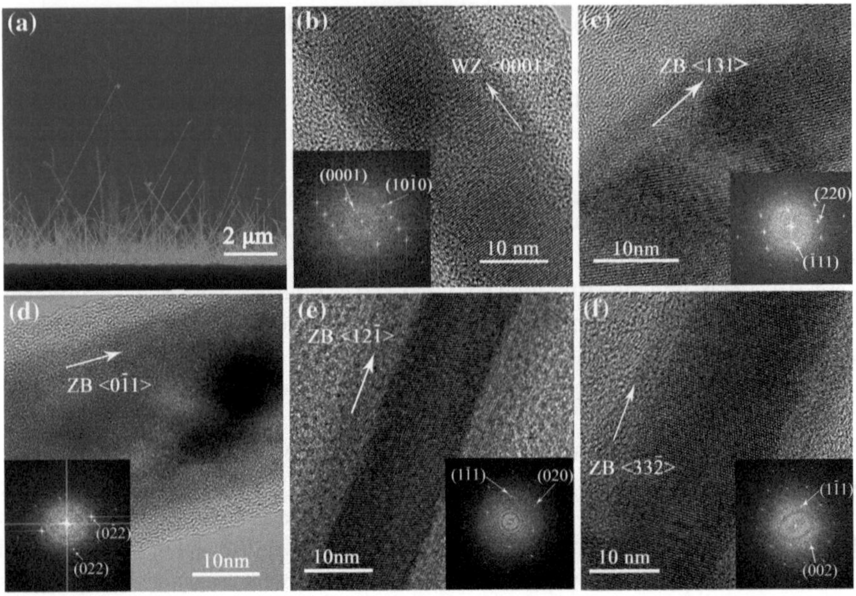

Fig. 4.6 **a** SEM image of the InAs NWs studied in this chapter. Five types of InAs NWs, including **b** <0001> oriented WZ, **c** <131>, **d** <011>, **e** <121>, and **f** <332> oriented ZB NWs, are observed by TEM. The thick amorphous layer outside the NWs is the amorphous carbon deposited during the TEM observation and nano-manipulation in SEM. HRTEM images show the single-phase structure of InAs NWs. The insets of (**b–f**) are the FFT patterns of the HRTEM. All these five types of InAs NWs are transferred from the FETs by nano-manipulation after their electrical properties have been characterized. Reprinted with the permission from Copyright 2016 American Chemical Society

In total we studied 21 devices and found five types of NWs, including <0001> oriented WZ NWs, <131>, <011>, <121> and <332> oriented ZB NWs. HRTEM study shows that most of the NWs investigated in this work are pure-phased single crystal with few stacking faults (Fig. 4.6).

4.4 Influence of Crystal Phase and Orientation on the Electrical Transport Properties of InAs Nanowires at Room Temperature

4.4.1 Comparison of Device Performance with Different Crystal Structure

First, we compare the electrical properties of the NWs with similar diameters (20 $\pm$ 2 nm) in order to avoid the influence of the diameter. As shown in Fig. 4.7, different transfer characteristics and output characteristics of the FETs are obtained

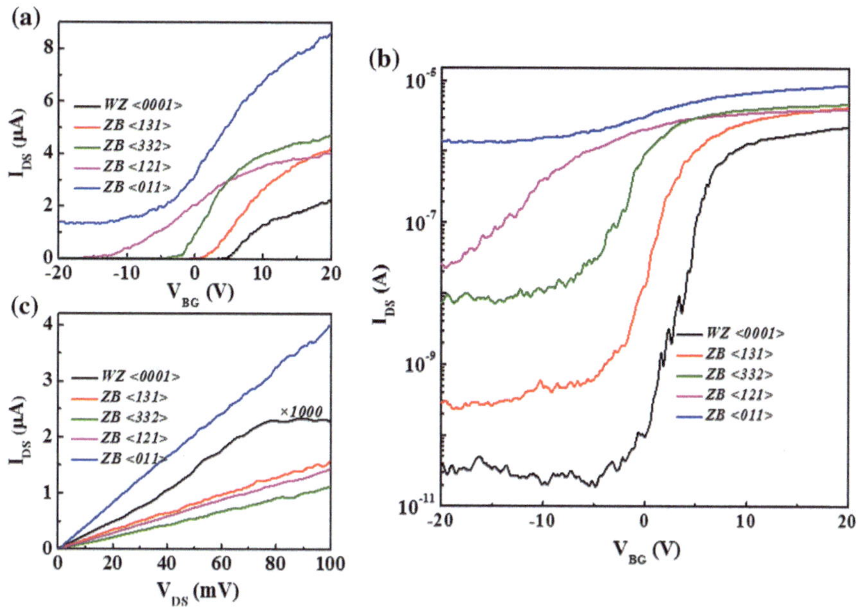

Fig. 4.7 $I_{DS} - V_{BG}$ characteristics at room temperature of the back-gated NW FETs at $V_{DS} =$ 0.1 V for five different oriented InAs NWs in the **a** linear and **b** logarithmic coordinates. **c** The I_{DS}–V_{DS} characteristics of these FETs. These five InAs nanowires have similar diameter. Reprinted with the permission from Copyright 2016 American Chemical Society

from five different types of NWs at the room temperature. It can be seen that both the crystal phase (WZ or ZB) and the orientation of the axis of the NWs affect the electrical transport properties. The V_T of the FETs based on the <011>, <121> and <332> oriented ZB InAs NWs are negative, while it is slightly above zero for the <131> oriented ZB NW, and the highest V_T is found in the <0001> oriented WZ NWs. According to I_{DS}–V_{DS} plots of InAs nanowires shown in Fig. 4.7c, ZB-phase InAs nanowires have much smaller resistance than WZ-phase InAs nanowires. Also, the *SS* of the FETs increases one by one in the sequence of <0001> oriented WZ NW, <131>, <332> and <121> oriented ZB NWs. The *SS* for the <011> oriented ZB NWs can not be reliably extracted because of their very small ON-OFF ratio. The ON-OFF ratio of the FETs is also influenced by both the crystal phase and the orientation as shown in Fig. 4.8a. Specifically, it is more difficult to pinch off ZB-phase InAs nanowires than WZ-phase nanowires.

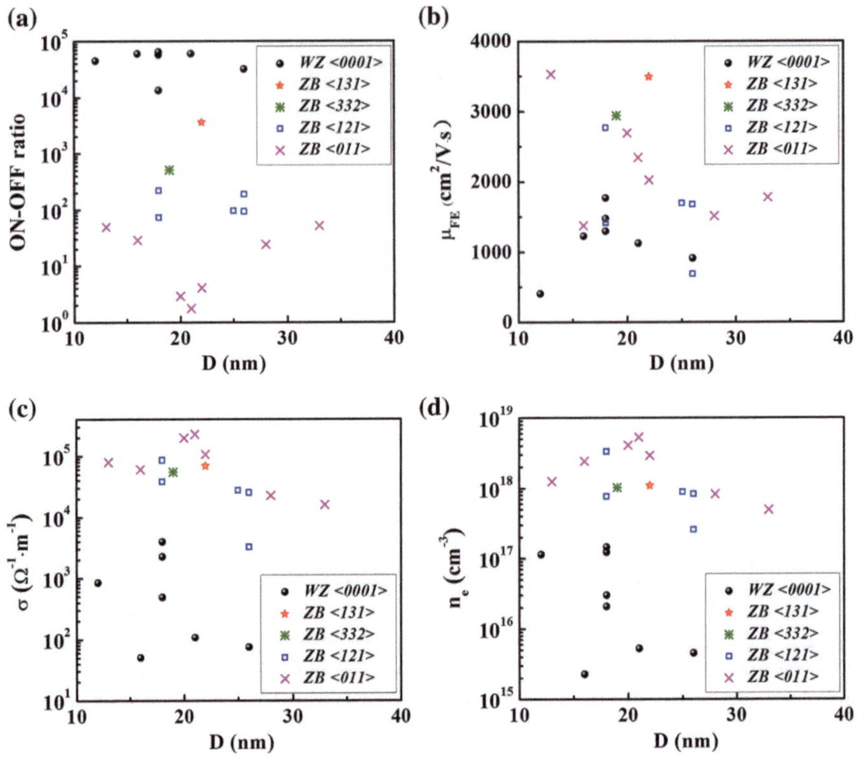

Fig. 4.8 The statistical data of several key electrical parameters, including the **a** ON-OFF ratio, **b** field-effect mobility, **c** conductivity and **d** effective electron concentration. These data were measured from 21 InAs NWs with various phases and orientations. Reprinted with the permission from Copyright 2016 American Chemical Society

4.4.2 Statistical Analysis on Device Parameters

We have a statistical analysis on the electrical properties of 21 devices. As shown in Fig. 4.8, the changes of the electrical transport properties with the crystal phase and orientation of the NWs are also observed in the FETs based on the InAs NWs with their diameters ranging from 12 to 33 nm. The present results show that compared with the crystal phase and the orientation, the effect of the diameter in this range is not obvious. The relatively small influence of diameter on performance of InAs nanowire FETs may result from the fact that the diameter is still large that a strong quantum confinement is not expected [16, 17].

As shown in Fig. 4.8a, the statistical ON-OFF ratio of back-gated devices are consistent with the devices in the previous section with a diameter of ~20 nm: the ON-OFF ratio of the FETs decreases in the order of <0001> oriented

WZ, <131>, <332>, <121> and <011> oriented ZB NWs. All the <0001> oriented NWs present high ON-OFF ratio above 10^4, possibly due to a larger E_g of WZ InAs NWs that is helpful to suppress the leakage current at the OFF-state. The OFF-state property of <0001> oriented InAs NWs indicates that the back gating used in our work is efficient enough to electrostatically control the channel. Among four types of ZB InAs NWs, only the <131> orientation has a relatively ON-OFF ratio on the order of 10^3. The other three types of ZB NWs with their axes along <332>, <121> and <011> show smaller values around 10^3, 10^2 and 10, respectively, each within a small spread of less than one order. Half of <011> oriented ZB NWs show ON-OFF ratio less than 5. These experimental results show clear distinction and correlation in the ON-OFF performance in InAs NWs with different crystal directions. This points to a more facile way to separate the crystal phase and orientation of the NWs based on the electrical properties rather than direct TEM which is in many cases not practical after device fabrication.

On the other hand, as shown in Fig. 4.8b–d, we find that the μ_{FE}, conductivity (σ) and effective electron concentration (n_e) show no discernible dependence on the crystal orientations, but are largely influenced by the crystal phase. As shown in Fig. 4.8b, The extracted μ_{FE} of <0001> oriented WZ NWs is less than 1800 cm^2/V s, while most of the ZB-structured NWs show higher electron mobility with the mean value being around 2200 cm^2/V s. It should be noted that we used a relative dielectric constant of 2.5 for the resist S1813 [18] to calculate μ_{FE}. As shown in Fig. 4.8c, d, both σ and n_e of the WZ NWs are several order of magnitude smaller than that of ZB NWs. The value of n_e of the four different ZB structured NWs at $V_{BG} = 0$ V is more than 10^{18} cm^{-3} (10^{12} cm^{-2} if normalized by the perimeter), which is much higher than the intrinsic carrier concentration of bulk ZB InAs (10^{15} cm^{-3} at room temperature) [19] and is comparable to the reported value of surface 2-dimensional electron gases (2DEGs) on InAs NWs [20]. This result indicates that high density of surface donor states near the bottom of the conduction band reported in bulk ZB InAs and <111> oriented ZB NWs may also exist in ZB phase NWs. And these states form a surface accumulation layer [21] of electrons, causing the Fermi level to be close to the conduction band bottom. Therefore, in ZB-structured InAs NWs, a negative V_{BG} is needed to deplete the electrons and pinch off the devices although the present InAs NWs are undoped during the MBE growth.

It has been suggested that higher growth temperature would favor the formation of ZB structure as well as an increased indium content. However, as the present NWs are picked from the same small area on the same piece of growth substrate, both the ZB and WZ NWs studied here are grown on the same temperature and in the same gas environment. Besides, we also study the composition of the NWs through X-ray energy dispersive spectrum (EDS) in TEM (as shown in Fig. 4.2). We find that the stoichiometry of both the WZ and ZB InAs NWs are the same within the accuracy of EDS analysis. Therefore, we can exclude the possibility that the difference on the electrical transport properties between ZB and WZ NWs is originated from different stoichiometry in the two crystal phases.

Previous work on MOCVD-grown InAs NWs with 80 nm in diameter suggests that WZ phase NWs have larger μ_{FE} than the ZB phase NWs at the low temperature [6]. While extended WZ segments in predominantly ZB phase NWs (thicker than 40 nm) have been observed to increase the resistivity significantly [5]. However, our present results on MBE NWs thinner than 33 nm indicate the ZB phase NWs have larger μ_e and smaller resistivity than the WZ phase NWs. Moreover, we notice that the electrical properties of InAs nanowires with diameter studied in this chapter show much less dependence on the variation of diameter than that of WZ InAs nanowires with diameter smaller than 10 nm in Chap. 3. Especially for nanowires with diameter larger than 20 nm, compared with the effect of crystal phase and crystal orientation, the diameter of the InAs NWs does not show obvious effect on the ON-OFF ratio, conductivity and electron concentration.

4.5 Influence of Crystal Phase and Crystal Orientation of InAs Nanowires on Electrical Transport Properties at Low Temperature

OFF-state performance and contact properties are important parameters for devices. To investigate the influence of the crystal phase and orientation on the OFF-state and contact properties, low temperature measurements are performed to extract the effective barrier height (Φ_B) and Schottky barrier heights (Φ_{SB}) at the Cr-contact area in InAs NWs with different orientations. We measure the transfer characteristics of <0001> oriented WZ, <131>, <121> and <011> oriented ZB NWs at variable temperatures as shown in Fig. 4.9a, d, g and j. In high-temperature region, the conventional thermionic emission theory is used to extract Φ_B [16, 22, 23]:

$$I_{DS} = A^{*}T^{2}\exp\left(-\frac{q\emptyset_{B}}{k_{B}T}\right)\left[1-\frac{\exp(-qV_{DS})}{k_{B}T}\right] \tag{4.1}$$

where A* is the Richardson's constant, k_B is the Boltzmann constant, q is the charge of an electron, T is the temperature and V_{DS} is the voltage bias across the device. We applied a V_{DS} of 0.2 V, which is higher than $3k_BT$, on the devices in order to exclude the effect of the drain electrode and to simplify the relation between ln (I_{DS}/T^2) and $1/T$. Corresponding Arrhenius plots for various V_{BG} values of the four types of NWs are shown in Fig. 4.9b, e, h and k. The effective potential barrier Φ_B can then be extracted from the slope of the linear parts of the plots at high-temperature region (generally T > 180 K in our work), and the $\Phi_B - V_{BG}$ relations are obtained and shown in Fig. 4.9c, f, i and l.

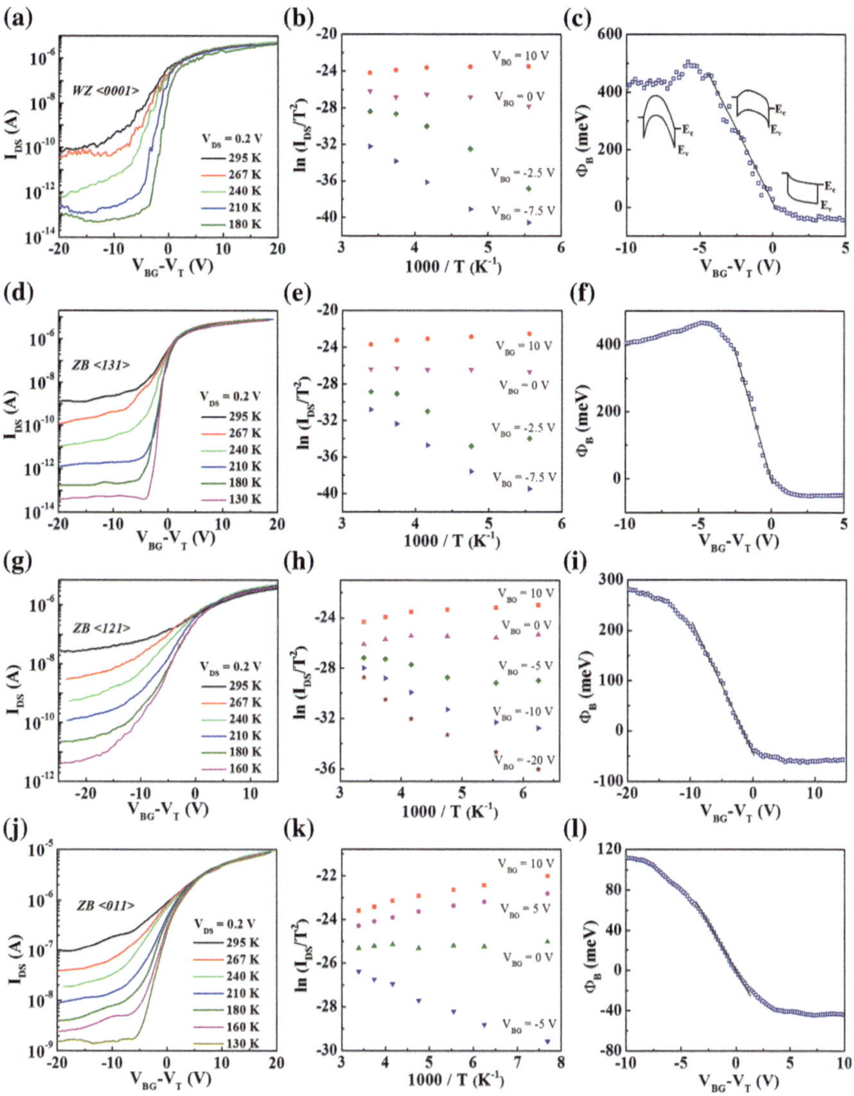

Fig. 4.9 (**a**, **d**, **g**, **j**) Temperature-dependent transfer characteristics of the back-gated FETs based on WZ <0001>, ZB <131>, ZB <121> and ZB <011> oriented InAs NWs, respectively. Corresponding Arrhenius plots for various gate voltages are shown in (**b**, **e**, **h**, **k**) and the extracted effective barrier height versus V_{BG} are plotted in (**c**, **f**, **i**, **l**). The inserted graph in (**c**) is the energy band diagram at different $V_{BG}-V_T$ conditions. Reprinted with the permission from Copyright 2016 American Chemical Society

4.5.1 *Influence of Crystal Phase and Crystal Orientation on the Contact Properties of InAs Nanowires*

As shown in Fig. 4.10, in the subthreshold region (when $V_{BG} \ll V_T$) of FETs, Φ_B follows a linear response to V_{BG} because the thermionic emission dominates the conduction mechanism [23]. The deviation of non-linear slope of Φ_B to V_{BG} from the ideal linear relation $\Delta\Phi_B = e\Delta V_G$ is the result of relatively thick gate dielectric layer of nanowire devices in this work. The flat-band voltage (V_{FB}) is extracted from the end of this linear relation to mark the transition from the thermionic emission region to the thermally assisted tunneling region when Φ_B is lower than Φ_{SB}. Because when $V_{FB} < V_{BG} < V_T$, assisted tunneling also plays an role in the conduction mechanism in the contact area, Φ_B deviates from the linear relation with V_{BG}, and Φ_B is lower than Φ_{SB}. We observe in Fig. 4.9 that both WZ and ZB NWs have V_{FB} very close to V_T, and the extracted Φ_B at V_{FB} (which is the Schottky barrier height at the contact) are all smaller than zero. It indicates ohmic contact at the Cr-InAs NWs contacts for all InAs nanowires studied in this chapter, regardless of crystal phases and orientations.

Bulk InAs has been well known to form ohmic contact with metals due to the fermi pinning at the bottom of conduction band [24]. Our work further proved that even without alloying process by annealing, Cr forms ohmic contact with the undoped InAs NWs with different crystal phases and crystal orientation. This ohmic contact between InAs nanowires and metal provides a big advantage in both traditional and novel devices.

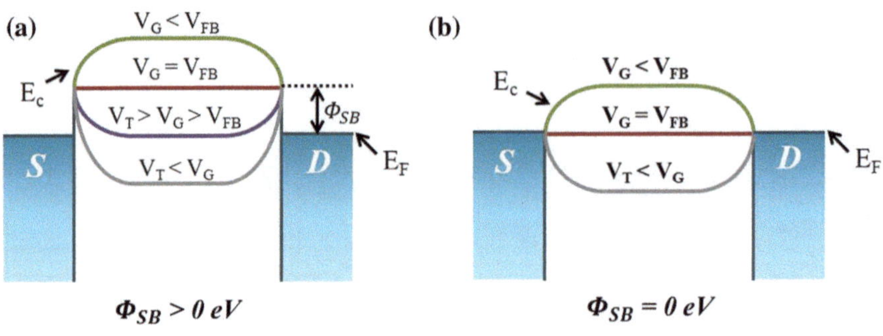

Fig. 4.10 Energy band diagram of InAs nanowire FETs at different gate voltages **a** when the contact between semiconductor nanowires and source/drain metal electrodes show Schottky barrier and **b** when semiconductor nanowires have ohmic contact with source/drain metal electrodes. Here, the valence band is omitted for clarity, since the electrons are majority carrier

4.5.2 Influence of Crystal Phase and Crystal Orientation of InAs Nanowires on the Barrier Height at OFF-State

When V_{BG} is turned to be more negative than V_{FB}, the effective barrier height Φ_B at the contact area gets higher and blocks the thermionic emission current, thus the conductance of the devices decreases and the current finally reaches the lowest value of the whole transfer curve. If a more negative V_{BG} is applied, the transfer curves show a plateau at the OFF-state or even a slight p-branch when the Fermi level aligns with or goes into the valence band, as shown in Fig. 4.9. In this region, the concentration of holes greatly increases and exceeds the concentration of the electrons in the channel. The hole tunneling current, rather than the electron current, plays the main role in the whole I_{DS}. From Fig. 4.9, we observe that the effective barrier extracted at negative V_{BG} does not increase continuously, but reaches a peak value and then decreases a little. The peak value of Φ_B in the Φ_B versus V_{BG} curve here should be the effective barrier at the OFF-state (Φ_{OFF}). Notably, the value of Φ_{OFF} follows the same tendency with the ON-OFF ratio. It monotonously increases from about 120 to 500 meV by the order of <011>, <121>, <131> oriented ZB and <0001> oriented WZ InAs NWs. High Φ_{OFF} can efficiently hinder the tunneling current flowing from the source to the drain electrodes so that suppresses the I_{off} and improves the ON-OFF ratio.

According to previous theoretical work, the Φ_{OFF} of ideal junctionless FETs based on one-dimensional narrow E_g nanomaterial is related to the E_g of the channel material and the Φ_{SB} at contact [25]. The ohmic contacts at the Cr-InAs NWs contact for electron transport and the very slight p-branch in the transfer curves in our work, the Φ_{OFF} is predicted to be closely related to the E_g of the InAs NWs. In the cases of <0001> oriented WZ NWs, the extracted Φ_{OFF} are generally equal to their E_g predicted by previous works [4, 26]. Moreover, that the Φ_{OFF} of the <0001> WZ InAs NW in our work is higher than all the ZB NWs is consistent with previous theoretical and experimental works which have verified that WZ structured InAs materials have larger band gap than ZB structured InAs [4, 7]. However, in the case of ZB InAs NWs, except <131> oriented ZB NWs, the other two types of ZB NWs both show Φ_{OFF} smaller than the E_g of 20 nm thickness ZB InAs NWs [9], which have been predicted theoretically to have no obvious distinction between different orientations.

4.6 Possible Mechanism for Influence of Crystal Phase and Crystal Orientation on the Electrical Properties of InAs Nanowires

Currently, the mechanism and impact of the crystal structure and orientation on the electrical properties of InAs nanowires is in controversy and lack systematic study. The diameter range studied in this chapter is mostly unexplored in previous theoretical calculations and experimental works. Therefore, our study in this chapter builds a connection between the existing theoretical and experimental results. In addition, because the research system used in this chapter excludes most of the non-ideal factors (such as background doping, crystal defects, etc.), our work has more guidance and reference significance for theoretical calculations.

4.6.1 Influence of Changes in the Band Structure

Many researchers have calculated the band structure of InAs in different crystal structure. It has been reported that WZ InAs bulk material have larger bandgap than ZB InAs bulk material by tens of meV to more than one hundred meV [4]. Moreover, the electron effective mass of WZ InAs bulk material is larger than that of ZB InAs bulk material, and a larger effective mass have been predicted along <0001> than orientation perpendicular to <0001> [4].

In our work, we observe very consistent results with these calculations. On the one hand, the increase of bandgap leads to higher barrier height at the contact at OFF-state, and thus suppresses thermal emission current and tunneling current. Consequently, lower I_{off} and higher I_{on}/I_{off} are generally obtained within WZ InAs nanowires. On the other hand, according to theoretical calculations, WZ InAs bulk material has electron effective mass about 0.060 m_0 in the direction of <0001>, which is much larger than the electron effective mass of ZB InAs bulk material (about 0.023 m_0). As a result, <0001> oriented WZ InAs nanowires show lower average field-effect electron mobility than the four kinds of ZB InAs nanowires in our work. However, we have also noticed that another theoretical simulation on the NWs thinner than 3 nm has suggested that <100> ZB NWs have larger electron effective mass than <110> ZB, <111> ZB, <0001> WZ and <10-10> WZ NWs, and the latter four types have about the same electron effective mass [10]. Presently, the InAs NWs in this chapter are 12–33 nm in diameter, beyond the scale calculated in theoretical works. Referring to our study on ultrathin nanowires in Chap. 3, the influence of quantum confinement effects in nanowires in this chapter are considered to negligible.

Regarding the influence of orientations on the bandstructure of InAs nanowires, theoretical calculations on the NWs have suggested that both the bandgap and the electron effective mass of will decrease by sequence of <100>, <111>, <110> when the diameter is in the range of 3–10 nm. When the diameter increases above 10 nm, the difference will disappear. Our work on MBE-grown InAs nanowires with diameter from 12 to 33 nm shows that crystal orientation has little influence on the electron field-effect mobility, and it is consistent with the above theoretical works on InAs nanowires and ZB InAs bulk material. It should be noted that different results have also been obtained in other calculations by different approximations and modeling methods [8]. In experiment, only Cui et al. reported the difference between the mobility of <011> -oriented ZB InAs nanowires and that of <111>-oriented ZB InAs nanowires when diameter of InAs nanowires is 40 nm [27]. However, we note that the <111>-oriented nanowires studied in Cui's work have large amount of stacking faults which might lead to the degradation of electron mobility, and thus it cannot support the claim that the growth direction is the main cause of different electron mobility. For the issue of whether the crystal orientation of InAs nanowires affects the effective electron mass, further theoretical and experimental efforts are required.

4.6.2 Influence of Changes in Surface Structures and Surface States

Though the difference in the band structure between WZ and ZB phase InAs can provide a good understanding of the differences shown in the ON-OFF ratio and the electron mobility of WZ and ZB InAs nanowires, however, it is not possible to well interpret the effect of the crystal orientations on the ON-OFF ratio of the nanowire-based FET devices according to the calculation results of band structures of ZB InAs nanowires with different crystal orientations. Therefore, we propose another possible explanation. The difference in surface properties of InAs nanowires with various orientations is another possible factor that causes the electrical properties to be related to their crystal structures and growth directions.

InAs nanowires have a very large surface to volume ratio, so that the property of surface is important to the band structure of the nanowires. Because <011>-oriented ZB InAs nanowires as well as <121>-oriented ZB InAs nanowires are likely to have the very imperfect surface or defects between nanowires and their oxides, it is likely that their band structure deviates from the ideal model in the theoretical calculations. The above assumption has been proved by a STM study [28]. The defects on the surface of <111>-oriented ZB InAs nanowires might cause the changes in measured band structure, and even lead to no bandgap in the bandstructure of InAs nanowires. In addition, by calculating the band structure of nanowires without passivating their surfaces by hydrogen, researchers have also confirmed that InAs nanowires passivated by different surface treatment have different band structure, and in some conditions, the InAs nanowires can show metallic properties rather than semiconductive properties [8]. Accordingly, we propose that the difference in growth direction of ZB InAs nanowires leads to their various surface states, thereby influencing their band structure and thus Φ_B at OFF-state. Moreover, a high density of surface states might induce high parasitic capacitance of surface which connects in series with the capacitance of the backgate [29] as shown in Fig. 4.11, and therefore reduce the equivalent capacitance of gate structure. Consequently, the reduced gate capacitance weakens the electrostatic coupling and leads to the variations of SS, ON-OFF ratio and Φ_B for different nanowire-based FETs. Also, this might account for the similar dependence of SS, ON-OFF ratio and Φ_B on crystal phases and crystal orientations as observed in Figs. 4.8 and 4.9.

In addition, the maximal effective electron concentration of <0001>-oriented WZ InAs nanowires does not exceed 10^{17} cm^{-3} which is lower than that of ZB InAs nanowires by one order of magnitude, while the minimal effective electron concen-

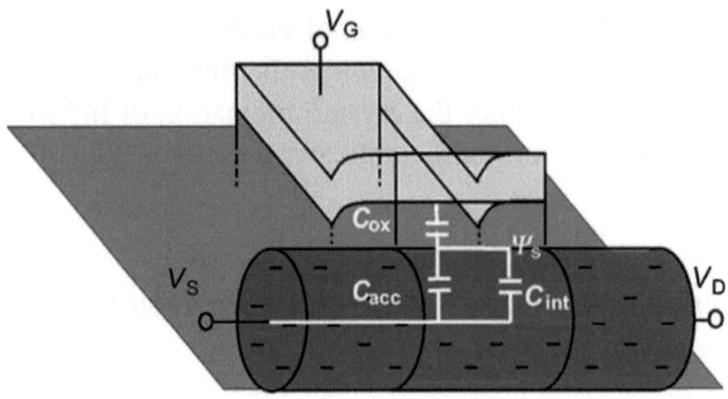

Fig. 4.11 An equivalent capacitance circuit diagram at the gate electrode of a nanowires device. In this figure, C_{ox} is the capacitance generated by gate oxide dielectric, C_{acc} is the capacitance introduced by depletion region or surface-accumulated carriers of semiconductor nanowires, and C_{int} is the interface capacitance induced by interface states between gate dielectric and semiconductor nanowires. Reproduced from ref. [29] by permission of John Wiley & Sons Ltd

tration of <0001>-oriented WZ InAs nanowires is around 10^{15} cm^{-3} [30], which is similar to the intrinsic electron concentration of bulk InAs material. Accordingly, we believe that the difference on the electron concentration is caused by the more perfect surface structure of <0001> WZ InAs nanowires than ZB InAs nanowires. We note that our experimental results are in accordance with measured bandstructure of InAs nanowire's surfaces in STM [7].

4.7 Summary

In this chapter, we developed an novel method for exploring the relation between the electrical properties and the crystal structure of InAs nanowires. This method combines an optimized nano-fabrication technique for planar suspended devices, nano-manipulation technique in SEM, and TEM characterization, so that we can obtain both nanowires' crystal structure at atomic level and device parameters for the same individual InAs nanowire.

To ensure a clean material system for studying crystal structure-dependent electrical properties of InAs nanowires, we used high-quality undoped InAs nanowires with uniform diameter grown in MBE chamber. We find that the electrical properties of InAs nanowire-based FETs are significantly impacted by both crystal phase and crystal orientation of nanowires. The *SS* increases while V_T, ON-OFF ratio and Φ_{OFF} decrease one by one in the sequence of WZ <0001>, ZB <131>, ZB <332>, ZB <121> and ZB <011> . While the WZ InAs NWs have obviously smaller field-effect mobility, conductivity and effective electron concentration at $V_g = 0$ V than the ZB InAs NWs, these three parameters are not sensitive to the orientation of the ZB InAs NWs. The diameter ranging from 12 nm to 33 nm shows much less effect than the crystal phase and orientation on the electrical properties of the InAs NWs.

Yet we observe that the good ohmic contact between InAs NWs and metal remains regardless of the variation of the crystal phase and orientation through temperature dependent measurements. Our work promotes deeper understanding of InAs NWs and is important for the development of nanowire-based devices.

References

1. Dick KA, Thelander C, Samuelson L et al (2010) Crystal phase engineering in single InAs nanowires. Nano Lett 10(9):3494–3499
2. Trägårdh J, Persson AI, Wagner JB et al (2007) Measurements of the band gap of wurtzite $InAs_{1-x}P_x$ nanowires using photocurrent spectroscopy. J Appl Phys 101(12):123701
3. Dayeh SA, Susac D, Kavanagh KL et al (2009) Structural and room-temperature transport properties of zinc blende and wurtzite InAs nanowires. Adv Func Mater 19(13):2102–2108
4. De A, Pryor CE (2010) Predicted band structures of III-V semiconductors in the wurtzite phase. Phys Rev B 81(15):155210
5. Thelander C, Caroff P, Plissard SB et al (2011) Effects of crystal phase mixing on the electrical properties of InAs nanowires. Nano Lett 11(6):2424–2429
6. Ullah AR, Joyce HJ, Burke AM et al (2013) Electronic comparison of InAs wurtzite and zincblende phases using nanowire transistors. Physica Status Solidi—Rapid Res Lett 7(10):911–914
7. Hjort M, Lehmann S, Knutsson J et al (2014) Electronic and structural differences between wurtzite and zinc blende InAs nanowire surfaces: experiment and theory. ACS Nano 8(12):12346–12355
8. Ning F, Tang L-M, Zhang Y et al (2013) First-principles study of quantum confinement and surface effects on the electronic properties of InAs nanowires. J Appl Phys 114(22):224304
9. Shimoida K, Yamada Y, Tsuchiya H et al (2013) Orientational dependence in device performances of InAs and Si nanowire MOSFETs under ballistic transport. IEEE Trans Electron Devices 60(1):117–122
10. Alam K, Sajjad RN (2010) Electronic properties and orientation-dependent performance of InAs nanowire transistors. IEEE Trans Electron Devices 57(11):2880–2885
11. Dick KA, Thelander C, Samuelson L et al (2010) Crystal phase engineering in single InAs nanowires. Nano Lett 10(9):3494–3499
12. Ullah AR, Joyce HJ, Burke AM et al (2013) Electronic comparison of InAs wurtzite and zincblende phases using nanowire transistors. Physica Status Solidi—Rapid Res Lett 7(10):911–914
13. Pan D, Fu M, Yu X et al (2014) Controlled synthesis of phase-pure InAs nanowires on Si(111) by diminishing the diameter to 10 nm. Nano Lett 14(3):1214–1220
14. Peng LM, Chen Q, Liang XL et al (2004) Performing probe experiments in the SEM. Micron 35(6):495–502
15. Wei X, Chen Q, Peng L et al (2010) In situ measurements on individual thin carbon nanotubes using nanomanipulators inside a scanning electron microscopy. Ultramicroscopy 110(3):182–189
16. Shi T, Fu M, Pan D et al (2015) Contact properties of field-effect transistors based on indium arsenide nanowires thinner than 16 nm. Nanotechnology 26(17):175202
17. Fu M, Pan D, Yang Y et al (2014) Electrical characteristics of field-effect transistors based on indium arsenide nanowire thinner than 10 nm. Appl Phys Lett 105(14):143101
18. Samiei E, Hoorfar M (2015) Systematic analysis of geometrical based unequal droplet splitting in digital microfluidics. J Micromech Microeng 25(5):055008
19. Sourribes MJ, Isakov I, Panfilova M et al (2014) Mobility enhancement by Sb-mediated minimisation of stacking fault density in InAs nanowires grown on silicon. Nano Lett 14(3):1643–1650

20. Blömers C, Grap T, Lepsa MI et al (2012) Hall effect measurements on InAs nanowires. Appl Phys Lett 101(15):152106
21. Olsson LO, Andersson CBM, Hakansson MC et al (1996) Charge accumulation at InAs surfaces. Phys Rev Lett 76(19):3626–3629
22. Appenzeller J, Radosavljević M, Knoch J et al (2004) Tunneling versus thermionic emission in one-dimensional semiconductors. Phys Rev Lett 92(4)
23. Razavieh A, Mohseni PK, Jung K et al (2014) Effect of diameter variation on electrical characteristics of Schottky barrier indium arsenide nanowire field-effect transistors. ACS Nano 8(6):6281–6287
24. Mead C, Spitzer W (1963) Fermi level position at semiconductor surfaces. Phys Rev Lett 10(11):471–472
25. Zhao Y, Candebat D, Delker C et al (2012) Understanding the impact of Schottky barriers on the performance of narrow bandgap nanowire field effect transistors. Nano Lett 12(10):5331–5336
26. Niquet Y, Lherbier A, Quang N et al (2006) Electronic structure of semiconductor nanowires. Phys Rev B 73(16)
27. Cui Z, Perumal R, Ishikura T et al (2014) Characterizing the electron transport properties of a single <110> InAs nanowire. Appl Phys Express 7(8):085001
28. Hilner E, Hakanson U, Froberg LE et al (2008) Direct atomic scale imaging of III-V nanowire surfaces. Nano Lett 8(11):3978–3982
29. Dayeh SA, Aplin DP, Zhou X et al (2007) High electron mobility InAs nanowire field-effect transistors. Small 3(2):326–332
30. Xu H, Wang Y, Guo Y et al (2012) Defect-free <110> zinc-blende structured InAs nanowires catalyzed by palladium. Nano Lett 12(11):5744–5749

Chapter 5
Influence of Different Growth Methods on the Electrical Properties of InAs Nanowires

Abstract In this chapter, we have studied the similarity and difference between the electrical properties of InAs nanowires grown by two commonly used material growth systems, MBE and MOCVD. Bases on the statistical data of more than 70 InAs nanowires back-gated FETs whose diameter range from 16 nm to more than 100 nm, we find that when the diameter of InAs nanowires is relatively small, most of the MOCVD-grown InAs nanowires have similar electron mobility, threshold voltage, ON-state current, and OFF-state current with MBE-grown InAs nanowires. However, the dispersion of these electrical properties within the MOCVD-grown InAs nanowires is much larger than that within the MBE-grown nanowires. On the other hand, when the diameter of InAs nanowires is relatively large, the ON-state properties of MBE- and MOCVD-grown nanowires does not show obvious difference, but their OFF-state properties have apparently different features. The MOCVD-grown nanowires have smaller OFF-state resistance than the MBE-grown nanowires, and some of the large-diameter nanowires even show metallic behavior. By simulating the distribution of electrons under ON-state and OFF-state in the nanowires with various doping levels through the finite element model, we attribute the background carbon doping induced by the organic sources used in the growing procedures of MOCVD to be the main cause for the above differences. Our work has deepened the understanding of the relation between nanowire growth and device performance, and may be instructive for controlling the growth of InAs nanowires.

Keywords Growth methods · Background carbon doping · OFF-state current ON-state current · Simulation of the distribution of electrons

In previous two chapters, we have systematically studied several device parameters that highlight the electrical properties of MBE-grown InAs nanowires. Due to the wide applications of MOCVD in industry, it is desired to compare the electrical properties of MBE- and MOCVD-grown InAs nanowires. It is known that background carbon doping is present in MOCVD-grown InAs nanowires, especially at higher growth temperature in order to avoid mixed ZB and WZ phases [1], which will change the electrical properties of grown nanowires. In this chapter, we study and compare the electrical properties of InAs nanowires grown by the above two methods.

M. Fu, *Electrical Properties of Indium Arsenide Nanowires and Their Field-Effect Transistors*, Springer Theses, https://doi.org/10.1007/978-981-13-3444-3_5

5.1 InAs Nanowires Grown by MBE and MOCVD

As shown in Fig. 5.1a, c, InAs nanowires grown by both MOCVD and MBE are perpendicular to the Si(111) substrate, and their growth directions are WZ <0001>. Through HRTEM characterization (Fig. 5.1b, d), it can be seen that the two kinds of InAs nanowires samples show similar crystal structure with the main WZ phase and a large number of stacking fault defects insides.

The InAs nanowire FETs invloved in this chapter are fabricated by the back-gated device process in Sect. 2.3.2. The source-drain electrodes are Cr/Au material and deposited by electron beam evaporation system, and before the deposition of metal, the native oxide layer of InAs nanowires are removed by $(NH_4)_2S_x$ solution. The diameter of InAs nanowires at the channel is characterized by AFM, and the

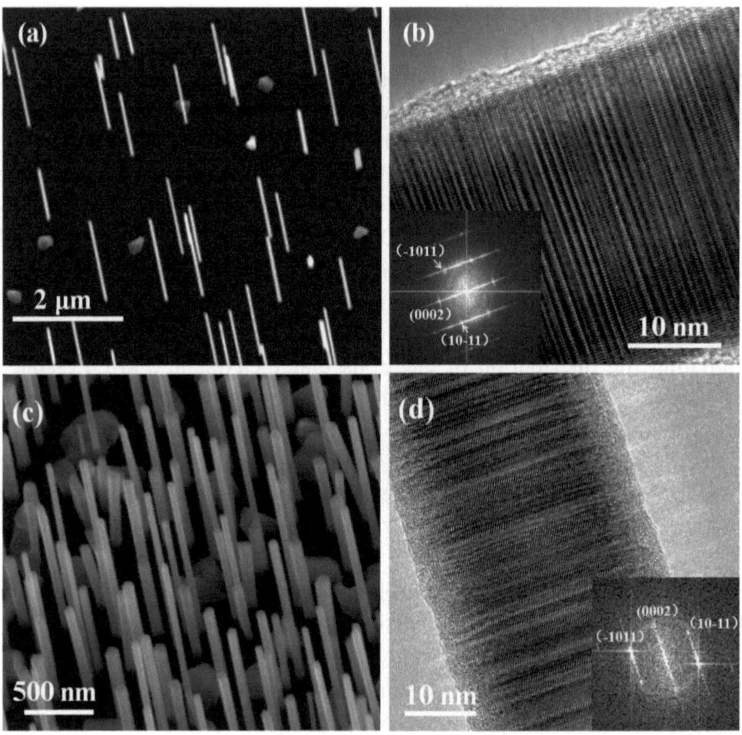

Fig. 5.1 Characterization of InAs nanowires self-catalyzed grown by MOCVD 与 MBE system. **a** SEM characterization of MOCVD-grown InAs nanowires. The grown nanowires are perpendicular to the Si(111) substrate; **b** TEM characterization of a MOCVD-grown InAs nanowire. The insert is the FFT of this TEM characterization. According to the FFT, the nanowire is mainly WZ phase with large number of stacking faults and grown along WZ <0001>. **c** SEM characterization of MBE-grown InAs nanowires; **d** TEM characterization of an MBE-grown InAs nanowire. The insert is the FFT of this TEM characterization. According to the FFT, the nanowire is mainly WZ phase with large number of stacking faults and grown along WZ <0001>

channel length of the device is measured by SEM. In order to prevent the devices from short-channel effects affecting the OFF-state characteristics of the device, the channel length of all devices is greater than 400 nm.

5.2 Comparison of Electrical Properties Between MBE and MOCVD Grown InAs Nanowires

In our work, a wide range of diameters are chosen, which is 22–110 nm for MOCVD nanowires and 16–95 nm for MBE nanowires, respectively.

5.2.1 Comparison Between the Device Performance of InAs Nanowires Grown by Different Growth Methods with the Same Diameter

Figure 5.2 shows a comparison between the output characteristic as well as the transfer characteristic of one MBE- and one MOCVD- grown nanowire device. The diameter of the two devices is 21 nm. The channel length is 450 nm and 1.4 μm, respectively. As shown in Fig. 5.2, these two InAs nanowires present similar OFF-state performance, including similar V_T and OFF-state current. By extracting the transconductance and field-effect electron mobility, we found that the electron mobility of these devices is similar. The MOCVD-grown InAs nanowires have field-effect electron mobility of 966 cm^2/V s, and MBE-grown nanowires have mobility of 892 cm^2/V s.

However, in large diameter nanowires, MBE-grown ones show better ON-OFF property than MOCVD-grown ones, as shown in Fig. 5.3. The diameter of these two nanowires in Fig. 5.3 are similar. The one grown by MBE has a diameter of 94 nm and a channel length of 500 nm, and the one grown by MOCVD has a diameter of 92 nm and a channel length of 1.5 μm. From the transfer curve of the devices (Fig. 5.3b), we find the MBE-grown nanowire shows smaller OFF-state current. The extracted field-effect mobility of MBE-grown nanowire (1019 cm^2/V s) is lower than that of MOCVD-grown nanowire (1753 cm^2/V s).

5.2.2 Relation Between ON-OFF Ratio and Diameter of InAs Nanowire FETs

Figures 5.4 and 5.5 show the ON-OFF ratio of MBE and MOCVD grown InAs nanowires as a function of diameter. We investigated in total 70 devices. It can be observed that the ON-OFF ratios of devices in both growth methods decrease

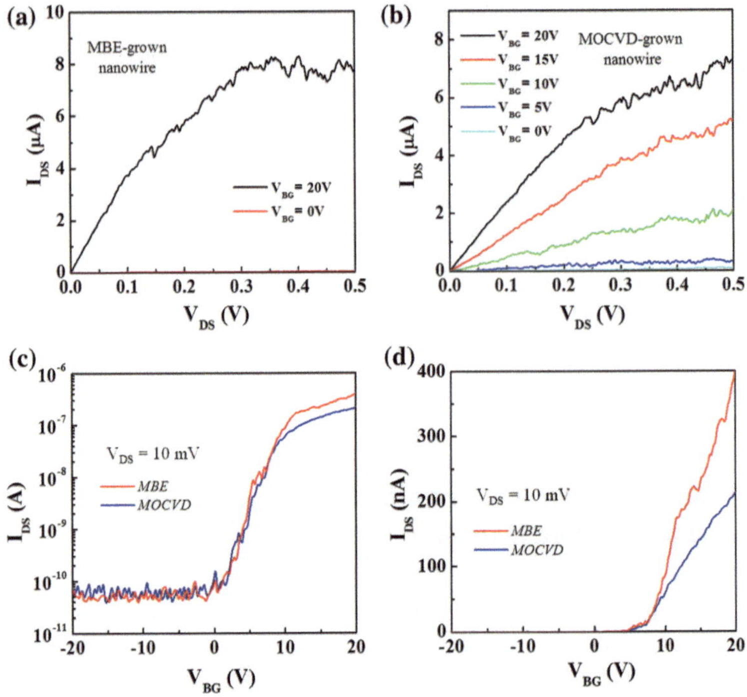

Fig. 5.2 The output characteristics and transfer characteristic of InAs nanowire back-gated FETs grown by different methods. The diameter of these nanowires is both 21 nm. The channel length of MBE-grown nanowire FET is 450 nm, and that of MOCVD grown nanowire FET is 1.4 μm. **a** Output characteristic of the MBE-grown InAs nanowire FET. **b** Output characteristic of MOCVD-grown InAs nanowire FET. Comparison between the transfer characteristics of these two InAs nanowires **c** in exponential coordinate and **d** linear coordinate

when the diameter increases. For MBE-grown nanowires, as shown in Fig. 5.4, the $\log(I_{on}/I_{off})$ of devices basically follows a parabolic drop as the diameter of nanowires changes. As the diameter becomes larger, the decreasing slope of $\log(I_{on}/I_{off})$ in small diameter region is larger than that in large diameter region. By linearly fitting the $\sqrt{\log(I_{on}/I_{off})}$与 to the diameter (as shown in the insert of Fig. 5.4), we find that the linear correlation coefficient R between $\sqrt{\log(I_{on}/I_{off})}$ and the diameter is 0.92. In the range of from 16 nm to 95 nm in diameter, the ON-OFF ratio of MBE-grown nanowires FETs decreases from 10^4 to 10, and then tend to slowly approach 1.

As for the MOCVD-grown InAs nanowires, the ON-OFF ratio of their back-gated FETs has much larger dispersion, as shown in Fig. 5.5, and thus the data points of the presented ON-OFF ratio in Fig. 5.5 are not able to be easily fitted by linear curve or parabolic curve. When the diameter of InAs nanowires is small (around 20–40 nm), the MOCVD-grown nanowires show similar $\log(I_{on}/I_{off})$~D relation with MBE-grown nanowires despite several deviation points in Fig. 5.5. However, when

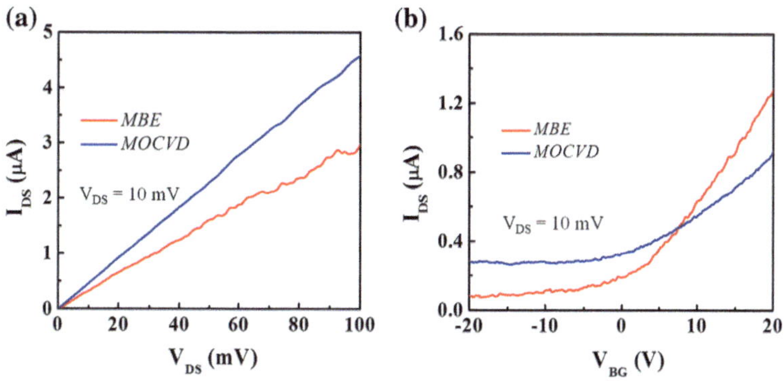

Fig. 5.3 The output characteristics and transfer characteristic of InAs nanowire back-gated FETs grown by different methods. The diameter of MBE-grown and MOCVD-grown nanowires are 95 and 92 nm, respectively. The channel length of MBE-grown and MOCVD-grown nanowires are 500 nm and 1.5 μm, respectively. **a** Comparison of output characteristics between these two InAs nanowires. **b** Comparison of transfer characteristics between these two InAs nanowires

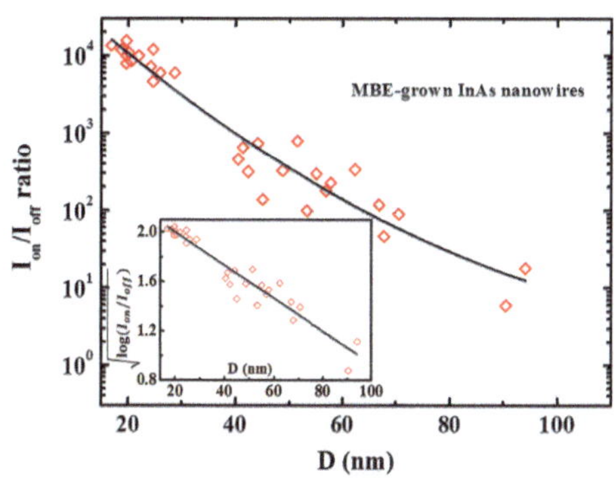

Fig. 5.4 Relation between the ON-OFF ratio of back-gated FET based on MBE-grown InAs nanowire and diameter of nanowires. As the diameter of nanowires increases, the ON-OFF ratio of nanowire FETs parabolically decreases. The insert shows the relation between $\sqrt{\log(I_{on}/I_{off})}$ of devices and diameter of nanowires

the diameter increases to above 60 nm, the MOCVD-grown nanowire FETs present weak ON-OFF characteristic; and when the diameter of nanowires continues to increase to above 80 nm, the ON-OFF ratio of MOCVD-grown InAs nanowire FETs is near 1, even if we apply a back-gate voltage ranging from +80 V to −80 V. This means that the MOCVD-grown InAs nanowires with diameter above 80 nm present metallic properties rather than the semiconductor's behavior.

As for the difference between MBE- and MOCVD- grown nanowires, according to transfer curves of nanowires grown by these two kinds of methods with the same diameter in last section, we consider the difference of the OFF-state current to be the major factor leading to the aforementioned difference on ON-OFF properties of InAs nanowires. And by comparing the electrical properties of pure-phase InAs

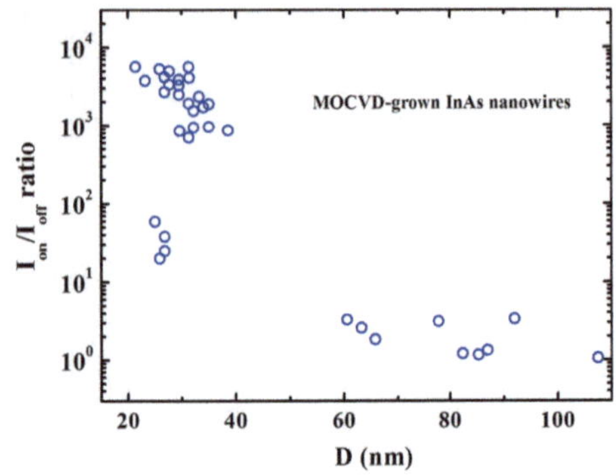

Fig. 5.5 The relation between ON-OFF ratio of back-gated FETs based on the MOCVD-grown InAs nanowires and the diameter of nanowires. As the diameter increases, the ON-OFF ratio of FETs parabolically decreases

nanowires and nanowires with large density of stacking faults, we exclude the effects of stacking faults on the OFF-state current. Below, we propose that the difference in electrical properties of MBE- and MOCVD- grown InAs nanowires may be caused by the doping of background carbon introduced during growth in MOCVD.

5.2.3 *Qualitative Understanding of the Effect of Growth Methods on the ON-OFF Ratio of InAs Nanowire FET*

Because the nanowires are not intentionally doped in the growth, we assume they are either undoped or uniform doped by the background carbon, and therefore the nanowire-based FETs are junctionless devices. According to the statements of junctionless devices in the Sect. 2.5 of Chap. 2, in order to pinch off the junctionless devices, the majority carriers in the channel are needed to be depleted. As for InAs nanowires, the majority carriers are electrons. According to classical theory of semiconductor physics, the depletion width (WD) of a semiconductor is related to surface potential, semiconductor dielectric constant, and doping concentration. Their relation can be expressed as [2]:

$$W_D = \left(\frac{2\varepsilon_{\mathrm{InAs}}\varphi_s}{eN_e}\right)^{0.5} \tag{5.1}$$

Here, as shown in Fig. 5.6, φ_s is the surface potential of the interface between InAs nanowire and the gate dielectric, ε_{InAs} is the dielectric constant of InAs nanowires, and N_e is the doping concentration of nanowires (electron concentration in intrinsic semiconductor). Because MBE-grown InAs nanowires have few dopants, we consider them as an intrinsic semiconductor, and thus their inner N_e ranges from 10^{15} to 10^{16} cm^{-3}. Accordingly, when the surface potential at the interface between InAs nanowires and SiO_2 is 0.1 eV, the width of the depletion region is about several tens

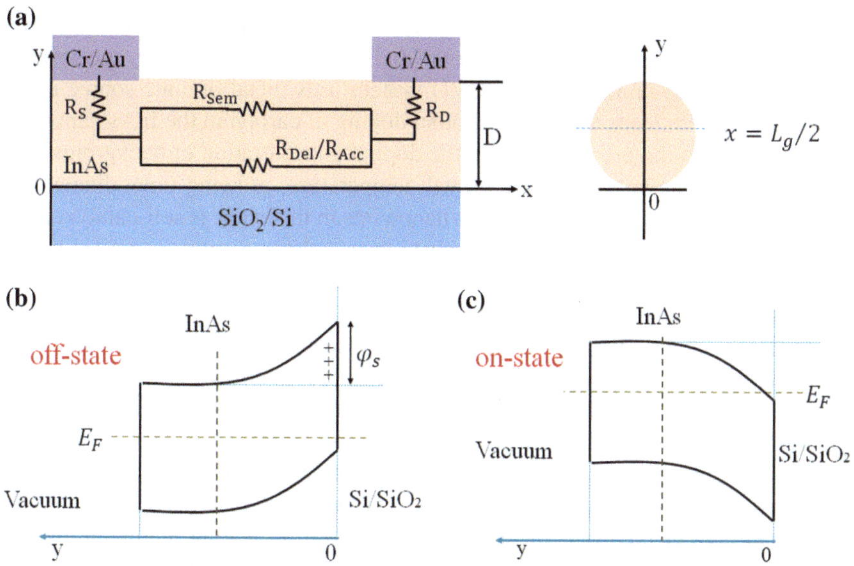

Fig. 5.6 **a** Schematic diagram of the resistance distribution of the InAs nanowires back-gated FETs. The band diagrams of the InAs nanowires at the OFF-state and the ON-state, from the surface (y = 0) to the inner body, are shown as (**b**) and (**c**), respectively. In this figure, it is assumed that the nanowires are only unilaterally regulated by the plate capacitor

to several hundreds of nanometers. Therefore, MBE-grown InAs nanowires with a diameter of approximately 90 nm can still be modulated by the back gate.

As shown in Fig. 5.6, if the interface between InAs nanowires and SiO_2 is set to the zero point of y-axis, and the combination of nanowires and Si/SiO_2 back gate is simplified to a plate capacitor model, then the potential at different y values can be obtained by the Poisson equation, and according to the following equation:

$$n_e = n_i \exp\left(\frac{E_c - E_F}{k_B T}\right) \tag{5.2}$$

the electron concentration (n_e) at different y values can be calculated, and then the number of electrons participating in the conduction in the nanowires can be obtained by integrating n_e. Wherein, n_i is the intrinsic electron concentration of InAs nanowire, E_c is the energy level at the bottom of conduction band of nanowire, E_F is the Fermi level of nanowire, k_B is the Boltzmann constant, and T is the temperature. Because compared with the InAs material at the interface between nanowires and SiO_2, the InAs material at the core (or upper part) of the nanowires is relatively weakly controlled by the gate, the electron concentration at the core of the large-diameter nanowires is much higher than that at the interface between nanowires and SiO_2. This results in a higher OFF-state current of large-diameter nanowire FETs than that of small-diameter nanowire devices.

In the situation of the MOCVD-grown nanowires, the carriers contributing to the conduction comprise both intrinsic carriers and electrons introduced by the unintentional doping. It is known that MOCVD system uses metal organic source as the InAs source which leads to the background doping of carbon in the InAs nanowires [3]. According to the previous works, the doping concentration of background carbon varies with the change on the growth temperature, growing time, diameter of nanowires etc. [3–5]. Because the InAs nanowires in this work is self-catalyzed and this self-catalyzed process need a relatively high growth temperature as well as hydrogen (which is the carrier gas), the background carbon doping of our MOCVD-grown nanowires might be lower than the nanowires grown in low-temperature conditions. Consequently, if the diameter of the self-catalyzed InAs nanowire is smaller than the maximum width of the depletion region, then the nanowire is still likely to be depleted by the electric field. This can explain why the ON-OFF ratio of the back-gated FETs of small-diameter InAs nanowires grown by these two kinds of methods show similar change rate and tendency as the diameter increases in Figs. 5.4 and 5.5.

However, when the scale becomes small, it gets difficult to reach uniform doping. Through simple calculation, we find that only about 30 dopant atoms are needed to convert the cylindrical intrinsic InAs nanowires with a diameter of 20 nm and length of 1 μm into the doped semiconductor with a concentration of 10^{17} cm^{-3}. As a result, the MOCVD-grown InAs nanowires with small diameter are more likely to exhibit fluctuations in the doping concentration. The maximum width of depletion region is sensitive to the variation of doping concentration, so we can observe that in Fig. 5.5 the ON-OFF ratio of back-gated FETs of MOCVD-grown InAs nanowires have large dispersion.

When the diameter of MOCVD-grown InAs nanowires is relatively larger, as shown in Fig. 5.7, though the electrons near the interface between nanowires and gate dielectrics are depleted at the very negative gate voltage, the depletion region will not extend into the core. As a result, the core of the nanowires still has the high concentration of movable electrons, and consequently the nanowire-based FETs cannot be pinched off. Moreover, it can also be seen in Fig. 5.7 that the large-diameter nanowires show more consistent OFF-state properties than the small-diameter nanowires, because the doping in the large-diameter InAs nanowires is more uniform. As reported in the previous works, the thicker InAs nanowires have higher conductivity than the thinner nanowires [6]. Thus the large-diameter MOCVD-grown InAs nanowires usually present weaker response to the modulation of gates.

In order to study the electrical properties of these two kinds of samples more quantitatively, we further carried out a finite element method to simulate the depletion region width and the carrier distribution of InAs nanowires with different doping concentrations.

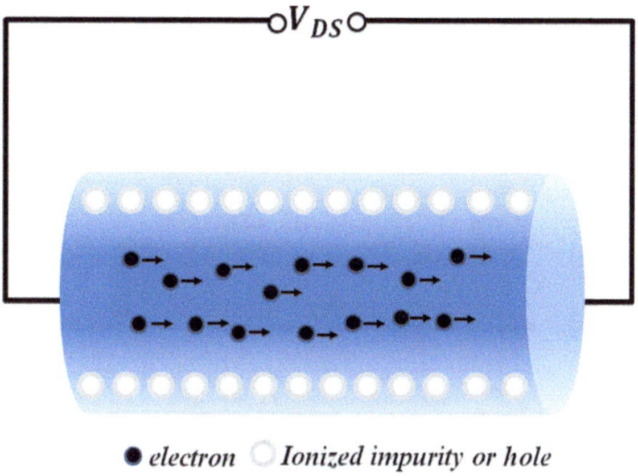

Fig. 5.7 The schematic diagram of carrier distribution of InAs nanowires device with a large diameter ($D > 2W_D$) at OFF-state. In this figure, an axially symmetric gate structure is used to simplify the model

5.2.4 Finite Element Simulation of InAs Nanowires with Different Doping Concentration

The finite element simulation platform used here is COMSOL MultiPhysics Finite-Element Modeling suite. In this model, the band structure of InAs is set to be similar with the bulk InAs. That is, the bandgap of InAs is set to be 0.36 eV. As shown in Fig. 5.6a, the device is simplified to be a two-dimensional plate capacitor model with a gate dielectric of 300 nm SiO_2 and a dielectric constant of 3.9. Since the transfer curve of InAs nanowires exhibits n-type characteristics in the experiment, the electrons are selected as majority carriers in the simulation, and only the transport properties of electrons are considered.

In this chapter, InAs nanowires with five kinds of doping concentrations were simulated: 5×10^{16}, 1×10^{17}, 5×10^{17}, 1×10^{18} and 5×10^{18} cm^{-3}. Wherein, the carrier distributions of InAs nanowires with doping concentrations of 5×10^{16}, 5×10^{17} and 5×10^{18} cm^{-3} at $V_{BG} = -20$ V are shown in Fig. 5.8a–c. As shown, when the doping concentration of nanowire is 5×10^{16} cm^{-3}, the width of depletion region of InAs nanowire can reach tens of nanometers. However, as the doping concentration increases, the width of depletion region decreases. As shown in Fig. 5.8d, if the doping concentration increase from 5×10^{16} to 1×10^{17} cm^{-3}, the width of depletion region will slightly decrease. But, the electron concentration on the surface of InAs nanowires barely has changes, and the distribution of electrons in these two kinds of nanowires have small difference in the region near the surface. Consequently, because the back-gate used in our devices have stronger electrostatic coupling with nanowires than the plate capacitor, and the back gate is able to modulate the upper surface of nanowires as well [7], thus the InAs nanowires having diameter of 20 nm show similar OFF-state performance as their doping concentration increase from 5×10^{16} to 1×10^{17} cm^{-3}. However, when the diameter of InAs nanowires increases to more than 100 nm, the electron concentration in their upper portions present a changes that is almost 1 order of magnitude and thus lead to the great difference on

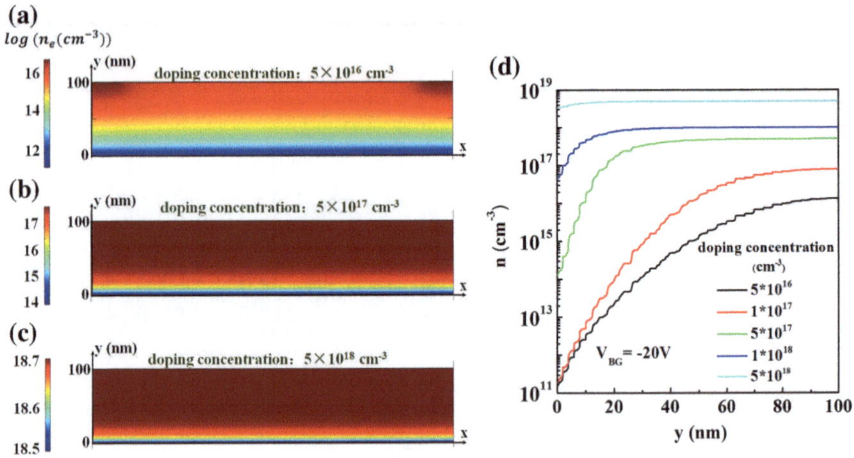

Fig. 5.8 **a–c** The cross-section diagrams showing the distributions of electron in the InAs nanowires with doping concentration of 5×10^{16}, 5×10^{17} and 5×10^{18} cm^{-3} when $V_{BG} = -20$ V; **d** The plots of the electron concentration of InAs nanowires with doping concentration of 5×10^{16}, 5×10^{17} and 5×10^{18} cm^{-3} when $V_{BG} = -20$ V versus the distance away from the interface between nanowire and dielectric

the OFF-state current. Further, when the doping concentration continues to increase to 1×10^{18} cm^{-3}, the width of depletion region will shrink to only a few nanometer. Meanwhile, the electron concentration at the interface between InAs nanowires and gate dielectrics becomes near 1×10^{17} cm^{-3} which is much higher than that (which is about 1×10^{11} cm^{-3}) of nanowires with doping level of 5×10^{16} and 1×10^{17} cm^{-3}. In this situation, even when the diameter of heavily-doped nanowires is less than 20 nm, the nanowire-based FETs might not be pinched off at the OFF-state. And with the diameter of nanowires increasing, these heavily-doped nanowires can show no response to the modulation of back-gate.

Although the doping concentration has a large effect on the OFF-state properties of InAs nanowire FETs, it has much less effect on the ON-state properties. As shown in Fig. 5.9, when the doping concentration of nanowires increases from 5×10^{16} to 5×10^{18} cm^{-3}, the electron concentration of the conductive layer on the surface is still at the same level at $V_{BG} = 20$ V. Specifically, when the doping concentration increases from 5×10^{16} to 1×10^{18} cm^{-3}, the electron concentration on the surface is not doubled. These simulation results are also consistent with the ON-state performance of InAs nanowires grown by two kinds of growth methods shown in Figs. 5.2 and 5.3.

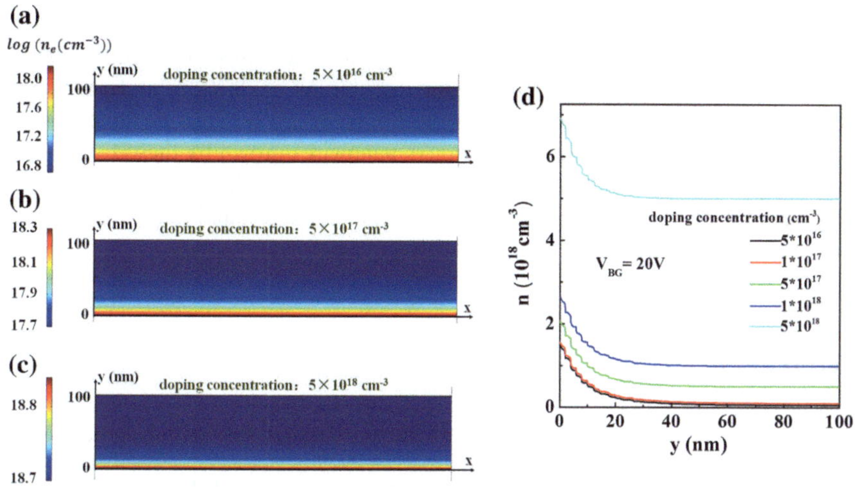

Fig. 5.9 **a–c** Cross-section of distribution of electron concentration in the body of nanowires at $V_{BG} = +20$ V, when the doping levels of InAs nanowires are 5×10^{16}, 5×10^{17} and 5×10^{18} cm^{-3}, respectively; **d** The plot of electron concentration in the body of InAs nanowires versus the distance away from the interface between nanowire and dielectric (y = 0), at $V_{BG} = +20$ V, when then doping levels of InAs nanowires are 5×10^{16}, 1×10^{17}, 5×10^{17}, 1×10^{18} and 5×10^{18} cm^{-3}, respectively

5.3 Summary

In this chapter, we have studied the similarity and difference between the electrical properties of InAs nanowires grown by two commonly used material growth systems, MBE and MOCVD. This study bases on the statistical data of more than 70 InAs nanowires back-gated FETs whose diameter range from 16 nm to more than 100 nm. And according to the transfer curves and output curves of these FETs, we find that when the diameter of InAs nanowires is relatively small, most of the MOCVD-grown InAs nanowires have some similar ON-state and OFF-state properties with MBE-grown InAs nanowires. These ON-state and OFF-state properties include electron mobility, threshold voltage, ON-state resistance, OFF-state resistance etc. However, the dispersion of these electrical properties within the MOCVD-grown InAs nanowires is much larger than that within the MBE-grown nanowires. On the other hand, when the diameter of InAs nanowires is relatively large, the ON-state properties of MBE- and MOCVD-grown nanowires does not show obvious difference, but their OFF-state properties have apparently different features. Specifically, the MOCVD-grown nanowires have smaller OFF-state resistance than the MBE-grown nanowires, and some of the large-diameter nanowires even show metallic behavior. However, regardless of the growth methods, the ON-OFF ratio of most of the InAs nanowires decreases with their diameter becoming larger.

We have attributed the background carbon doping induced by the organic sources used in the growing procedures of MOCVD to be the main cause for the difference of observation. Meanwhile, by simulating the distribution of electrons under ON-state and OFF-state in the nanowires with various doping levels through the finite element model, the electrical characteristics of InAs nanowires grown by these two growth methods are reasonably explained. Our work has deepened the understanding of the relation between nanowire growth and device performance, and may be instructive for controlling the growth of InAs nanowires.

References

1. Thelander C, Caroff P, Plissard SB et al (2011) Effects of crystal phase mixing on the electrical properties of InAs nanowires. Nano Lett 11(6):2424–2429
2. Sze SM, Ng KK (2007) Physics of semiconductor devices. Wiley, Hoboken
3. Larsen C, Li S, Buchan N et al (1990) Kinetics of the reaction between trimethylgallium and arsine. J Cryst Growth 102(1–2):126–136
4. Dick KA, Thelander C, Samuelson L et al (2010) Crystal phase engineering in single InAs nanowires. Nano Lett 10(9):3494–3499
5. ReepS DH, Ghandhi SK (1983) Deposition of GaAs epitaxial layers by organometallic CVD temperature and orientation dependence. J Electrochem Soc 130(3):675–680
6. Scheffler M, Nadj-Perge S, Kouwenhoven LP et al (2009) Diameter-dependent conductance of InAs nanowires. J Appl Phys 106(12):124303
7. Khanal D, Wu J (2007) Gate coupling and charge distribution in nanowire field effect transistors. Nano Lett 7(9):2778–2783

Chapter 6
Summary and Outlook

Abstract In this chapter, our researches of the influence of the size, the crystal structure and the growth method on electrical properties of InAs nanowires are summarized. Also, the further researches on nanowire-based devices are forecasted.

In this work, on the basis of InAs nanowire FETs, we study the affects of various factors, such as crystal structure, size, and growth method of InAs nanowires, on the electrical properties of InAs nanowires and their FETs. The main results are as follows:

1. The effects of size on the electrical properties of InAs nanowires is studied. When the diameter of InAs nanowires used as channel is smaller than 12 nm, because of the quantum confinement effect as well as the strong electrical coupling between gate and channel, the I_{on}/I_{off} ratio increases drastically with decreasing diameter of the InAs NWs, and when the diameter of the InAs NW is 7 nm, it can reach above 10^8, which is much higher than all the reported works of InAs nanowires. The threshold voltage also increases as the NW's diameter decreases. However, scaling down also brings severe surface scattering, decreased field-effect mobility, increased contact resistance etc.
2. The effect of the crystal phase and orientation on the electrical properties of InAs nanowires is studied. A novel method is developed to transfer the InAs NW from the channel of the device to the TEM, which allows atomic level structural characterization after electrical characterization. With this method, the electrical properties of single crystal InAs nanowires of five different crystal structures including the crystal phase and the crystal orientation are obtained. Parameters such as the threshold voltage, the ON-OFF ratio, the subthreshold swing and the effective barrier height at the OFF-state (Φ_{OFF}) are found to be greatly affected by both WZ and ZB phase and also the orientation of the NWs. The SS increases while V_T, ON-OFF ratio and Φ_{OFF} decrease in the sequence of WZ <0001>, ZB <131>, ZB <332>, ZB <121> and ZB <011>. The WZ InAs NWs have much smaller field-effect mobility, conductivity and electron concentration at

M. Fu, *Electrical Properties of Indium Arsenide Nanowires and Their Field-Effect Transistors*, Springer Theses, https://doi.org/10.1007/978-981-13-3444-3_6

$V_{BG} = 0$ V than the ZB InAs NWs, while these parameters are not sensitive to the orientation in the case of ZB NWs.

3. The effect of growth method on the electrical properties of InAs nanowires is studied. We find that when the diameter of InAs nanowire is relatively small, most devices based on MOCVD grown nanowires have similar ON- and OFF-state properties with devices of MBE-grown InAs nanowires, but the dispersion of OFF-state properties is much larger; on the other hand, when the nanowire diameter becomes large, the devices of both kinds show no significant difference in ON-state properties. The OFF-state resistance of MBE-grown nanowire are significantly larger than that of MOCVD-grown nanowires. Regardless of MBE or MOCVD grown InAs nanowires, their ON-OFF ratio of devices keeps decreasing with increasing diameter. By comparing the characteristics of these two growth methods, we believe that the background carbon doping introduced during the MOCVD growth might be the main reason of the differences in the electrical properties, and this is further supported by a finite element simulation of the distribution of electron concentration of nanowires having different doping concentrations.

The above research results contribute to the understanding and application of InAs nanowire based high-performance electronics. It should still be noted that a lot of problems are still open for future studies. For example, chemical or physical surface passivation process can reduce the density of surface states of InAs nanowires; how to further reduce the contact resistance between ultrathin InAs nanowires and metals; what is the cause of the huge difference in the electrical properties of InAs nanowires with different orientations etc.

Appendix
Experimental Parameters for Device Preparation

A.1 Locating Nanowires

In order to select the appropriate nanowires and deposit electrodes at the desired positions, the nanowires are required to be evaluated and precisely located. Therefore, the precise marks are needed to locate the nanowires and reduce the errors on nanowire locating and electrode exposing. In our works, these marks are usually made of metal film which contains of Au, such as Ti/Au or Cr/Au film. That is because: on the one hand, nanowires are small in size and thus are often needed to be located by SEM, while the Au has high contrast in SEM; on the other hand, in electron beam exposure process, the contrast of the Au film is also the highest when performing the manual mark correction.

Figure A.1 shows one of the commonly used patterns of small metal marks. The metal marks consist of two squares of 1×1 μm, and are better for determining their center position in electron beam exposure. Usually, every 500 μm write field have 5×5 marks. The lateral and longitudinal distance between two adjacent marks are of 40 and 30 μm, respectively. In environmental scanning electron microscopy, the magnification of approximately 6000 times can be used to put 4 small marks into the field of view during photograph. With this type of marks, the accuracy of locating nanowires is approximately 100–200 nm, and the minimum electrode line that can be exposed is approximately 150–200 nm. When fabricating the finer lines or more precisely locating the nanowires, the locating marks are needed to be more precise and the write field is needed to be smaller in exposure process.

Key points in mark fabrication

(a) The size of write field used to make marks is the same as that of write field used to expose device electrodes;
(b) The mark used to locating nanowires and the mark used to expose the device electrode need to be a set of marks.

M. Fu, *Electrical Properties of Indium Arsenide Nanowires and Their Field-Effect Transistors*, Springer Theses, https://doi.org/10.1007/978-981-13-3444-3

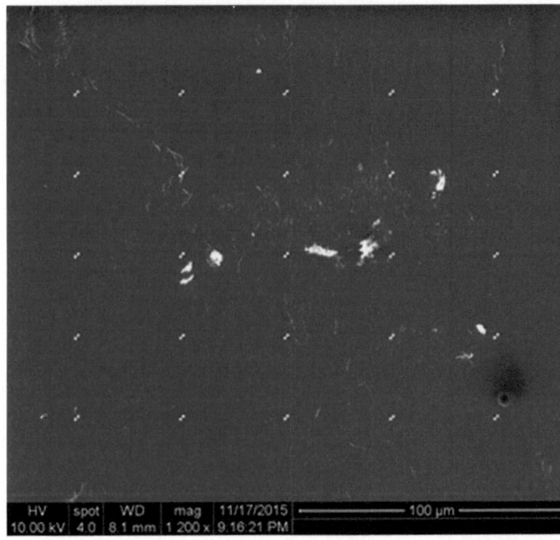

Fig. A.1 SEM image of InAs nanowires in the process of locating. The required nanowires have a diameter of approximately 20 nm

When fabricating marks, first the origin of exposure is determined, then the substrate is leveled, and after the writing field of exposure is corrected, the exposure is then directly performed. It is important to note that even if the large-size marks have been fabricated by photolithography or electron beam exposure on the substrate, it is not necessary to use these marks to do three-point correction, otherwise the accuracy of the locating coordinates will be lowered.

A.2 Treatment on Surface Oxide Layer of InAs at Source-Drain Contact

InAs are easily oxidized, so as shown in Fig. A.2, the oxide layer on the surface of InAs nanowires needs to be removed before vapor deposition of the source and drain electrode metal. In the fabrication process of planar devices, this oxide layer are usually etched by $(NH_4)_2S_x$ solution which is a mixture of ammonium sulfide solution (concentration of about 40%) and elemental sulfur (semiconductor grade). In addition to the $(NH_4)_2S_x$ solution, HF and HCl can also remove the oxide layer of InAs. Here, we use the $(NH_4)_2S_x$ solution mainly because of its surface passivation effect on InAs material. In addition, its toxicity and operational safety are relatively higher than HF.

The method to prepare the $(NH_4)_2S_x$ stock solution is as following:

(a) Mix 50 mL of $(NH_4)_2S$ solution with concentration of 40% with 4.8 g sulfur pellets (semiconductor grade) in a sealed bottle;

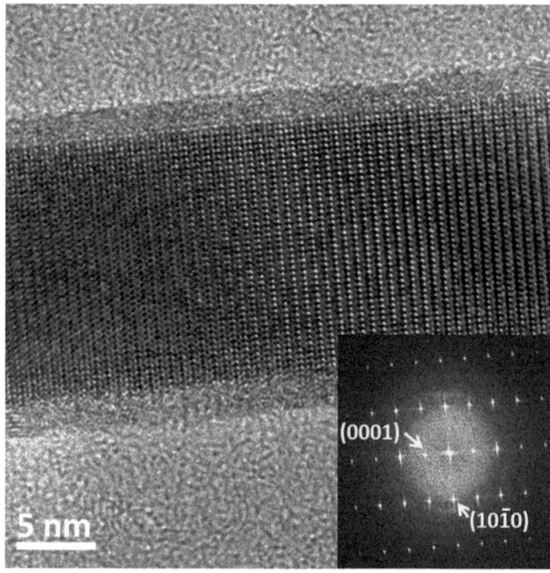

Fig. A.2 TEM image of a WZ phase InAs nanowire grown along <0001> direction

(b) Put a magnetic stir bar into the bottle and tight the lid of bottle, and then place the whole bottle on the magnetic stirring apparatus;
(c) After stirring the solution for 24 h, put the solution in the refrigerator for 24 h.

After the $(NH_4)_2S_x$ stock solution is prepared, it can be saved in the refrigerator for about half a year and then taken by a pipette when it is needed in experiments. The solution used to etch the native oxide layer of InAs nanowires is the diluted $(NH_4)_2S_x$ solution with a ratio of $(NH_4)_2S_x$ stock solution to deionized water =1:20 (volume ratio), and the etching time is about 1 min. The conventional reaction temperature is usually 40 °C in water bath. However, this reaction temperature has recently been found to be slightly higher and is easy to cause the PMMA 950 K to be denatured and fall off. Therefore, in the etching experiments, the reaction temperature can be appropriately lowered. It should be noted that the etching rate of $(NH_4)_2S_x$ solution increases and its stability decrease when the solution is stored for a long time. Therefore, it is recommended to use the as-prepared solution or the solution that is stored for less than half a month to treat the ultrathin InAs nanowires that require precise control on the reaction rate of etching. Another problem needed to be notice is that, due to the large surface tension of the aqueous solution, the hollows or voids between PMMA and substrate should be avoided after electron beam exposure (as shown in Fig. A.3), otherwise PMMA is easy to collapse onto the Si substrate.

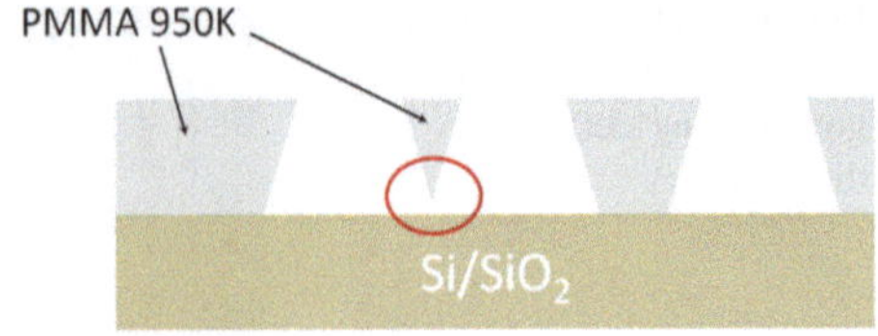

Fig. A.3 Hollows in PMMA under overexposure

A.3 ALD Deposition of High-k Dielectric Layers

In this thesis, two types of high-k dielectric layers are deposited by the Cambridge brand ALD: HfO_2 and Al_2O_3. Here, the condition for growing HfO_2 is introduced and its key parameters are as following:

(a) The growth temperature is 90 °C;
(b) At the beginning of growth, two Hf pulse is necessary before introducing the first H_2O pulse;
(c) Other specific growth parameters such as pulse time and wait time are adjusted according to the state of the device.

FSC
www.fsc.org
MIX
Papier aus verantwortungsvollen Quellen
Paper from responsible sources
FSC® C105338

If you have any concerns about our products,
you can contact us on
ProductSafety@springernature.com

In case Publisher is established outside the EU,
the EU authorized representative is:
Springer Nature Customer Service Center GmbH
Europaplatz 3, 69115 Heidelberg, Germany

Printed by Libri Plureos GmbH
in Hamburg, Germany